Electromagnetic Propagation and Waveguides in Photonics and Microwave Engineering

Edited by Patrick Steglich

Published in London, United Kingdom

IntechOpen

Supporting open minds since 2005

Electromagnetic Propagation and Waveguides in Photonics and Microwave Engineering
http://dx.doi.org/10.5772/intechopen.87925
Edited by Patrick Steglich

Contributors
Vladimir Schejbal, Ondrej Fiser, Vadim Zavodny, Helena Cristina Vasconcelos, Hedi Sakli, Wyssem Fathallah, Shotaro Ishino, Narumi Yashiro, Satoshi Denno, Seiichi Suzuki, Hong Lei Xu, Ting Ting Gu, Victor Mikhailov, Duy-Tien Le, Trung-Thanh Le, Vladimir Egorov, Evgeniy Egorov, Ran Gao, Jiansen Ye, Patrick Steglich

First published in London, United Kingdom, 2020 by IntechOpen
IntechOpen is the global imprint of INTECHOPEN LIMITED, registered in England and Wales, registration number: 11086078, 5 Princes Gate Court, London, SW7 2QJ, United Kingdom
Printed in Croatia

British Library Cataloguing-in-Publication Data
A catalogue record for this book is available from the British Library

Additional hard and PDF copies can be obtained from orders@intechopen.com

Electromagnetic Propagation and Waveguides in Photonics and Microwave Engineering
Edited by Patrick Steglich
p. cm.
Print ISBN 978-1-83968-188-2
Online ISBN 978-1-83968-189-9
eBook (PDF) ISBN 978-1-83968-190-5

For EU product safety concerns:
IN TECH d.o.o., Prolaz Marije Krucifikse Kozulić 3, 51000 Rijeka, Croatia, info@intechopen.com or visit our website at intechopen.com.

Meet the editor

Patrick Steglich is a research associate at the IHP - Leibniz-Institut für innovative Mikroelektronik, Germany, and lecturer for photonics and optical technologies at the Technical University of Applied Sciences Wildau, Germany. He obtained a master's degree in Photonics from the Technical University of Applied Sciences Wildau in 2013. In 2017, he received his PhD in Industrial Engineering from the Università degli Studi di Roma "Tor Vergata" for his work in the field of integrated photonics for communication and sensing. His research focuses on emerging photonic devices and waveguide concepts for telecommunication and sensing applications.

Contents

Preface

Optical and microwave waveguides have attracted much research interest in both science and industry. The number of applications for their use is rapidly growing. This book presents a collection of recent advances in the broad field of waveguide technology. It covers the current progress and latest breakthroughs in emergent applications in photonics and microwave engineering.

The book includes ten contributions on recent developments in waveguide technologies including theory, simulation, and fabrication of novel waveguide concepts.

The editor Patrick Steglich provides a brief overview of important developments in the field of waveguide technologies in photonics and microwave engineering as well as current trends and perspectives.

Ran Gao and JianSen Ye provide a review on optical fiber sensors based on anti-resonant reflecting optical waveguides, including the single layer, double layers, double resonators, and hybrid mechanism.

Helena Cristina Vasconcelos focuses on developing coatings for use as waveguides for integrated optics and photonics. Here, the waveguides are based on sol-gel materials.

Trung-Thanh Le and Duy-Tien Le present their results on graphene-silicon multimode waveguides. There results show that such waveguides can provide a very high free spectral range (FSR) and capability of controlling the critical coupling.

Hedi Sakli and Wyssem Fathallah present an extension of the rigorous analysis of the propagation of electromagnetic waves in transverse magnetic (TM) and transverse electric (TE) modes in a metallic circular waveguide partially filled with anisotropic metamaterial.

Shotaro Ishino, Satoshi Denno, Narumi Yashiro, and Seiichi Suzuki report on the development of a waveguide-type leaky-wave antenna and the development of a wireless LAN environment in a tunnel.

Vladimir Schejbal, Ondrej Fiser, and Vadim Zavodny investigate the radio-wave propagation above irregular ground, including the troposphere, using physical optics computation.

Ting Ting Gu, Hong Lei Xu, and Kai Li illustrate the disturbing nature during sunrise and sunset for Very Low Frequency (VLF) waves with periodic variations in amplitudes interfered by multi-modes over long propagation paths.

Egorov Evgenii Vladimirovich, and Egorov Vladimir Konstantinovich discuss waveguide-resonance mechanism relevation forerunner of characteristic X-ray radiation flux propagation and the technology of the planar air extended slit clearance preparation.

Viktor F. Mikhailov provides a solution for obtaining an analytical description of the radiation characteristics of a circular waveguide closed by a flat homogeneous dielectric plate.

We thank all authors for their contributions.

Finally, we express a great deal of thanks to the editing staff of IntechOpen, particularly Ms. Dolores Kuzelj, for all their efforts.

Sincerely,

Dr. Patrick Steglich
IHP – Leibniz-Institute for Innovative Microelectronics,
Frankfurt, Germany

Technical University of Applied Sciences Wildau,
Wildau, Germany

Section 1

Introduction

Chapter 1

Introductory Chapter: Electromagnetic Propagation and Waveguides in Photonics and Microwave Engineering

Patrick Steglich

1. Introduction

Waves can propagate as spherical waves in open space. In this case, the power of the wave decreases with the distance from the source as the square of the distance. In contrast, a waveguide can confine the propagating wave in such a way that the wave propagates only in one dimension. Assuming ideal conditions, the wave is not losing power while it propagates inside the waveguide.

Waveguides play a major role for applications in communications and sensing technologies. The theoretical understanding and practical investments are crucial to develop future innovations.

In photonics, two major types of waveguides can be distinguished, namely optical fibers and integrated waveguides. Waveguides in photonics operate typically in the visible and infrared light spectra.

Optical fibers are used for data transmission, as fiber lasers, for flexible transmission of laser radiation or for lighting, for sensor applications or decoration purposes [1]. The main application of optical fibers, however, is their use in telecommunication systems, making our daily life easier by a fast internet connection [2]. Other important technical applications of optical fibers are lasers [3], interferometers [4, 5], amplifier [6], and sensors [7]. The latter is important since it allows the detection of magnetic fields [8], humidity [9], temperature [10], and biological molecules [11, 12]. Massive research investments in the field of optical fibers [13–15] have led to novel applications. One important example is the use of optical for endoscopic applications [16, 17]. Also the fiber core has been modified (**Figure 1**), so that novel applications such as gas sensor can be addressed.

Integrated waveguides confine light in submicrometer structures on chip. Such waveguide structures are made either by doping the substrate material or by structuring it with etching procedures. Mostly, such waveguides are formed by patterning semiconductor materials like silicon, which is known as photonic integrated circuit technology [18]. The dimension of those waveguides in single mode operation is typically about 220 nm in width and 500 nm in height. **Figure 2** shows three different types of waveguides based on silicon.

The main applications are electro-optical modulators in telecommunications [19] and integrated sensors [20, 21] for point-of-care-diagnostics, environmental monitoring, or food analysis [22, 23]. A relatively novel approach is the silicon-organic hybrid technology [24–26]. Here, the silicon-based waveguide is covered with organic materials [27–29] leading to highly energy-efficient modulators [30] with large 3-dB

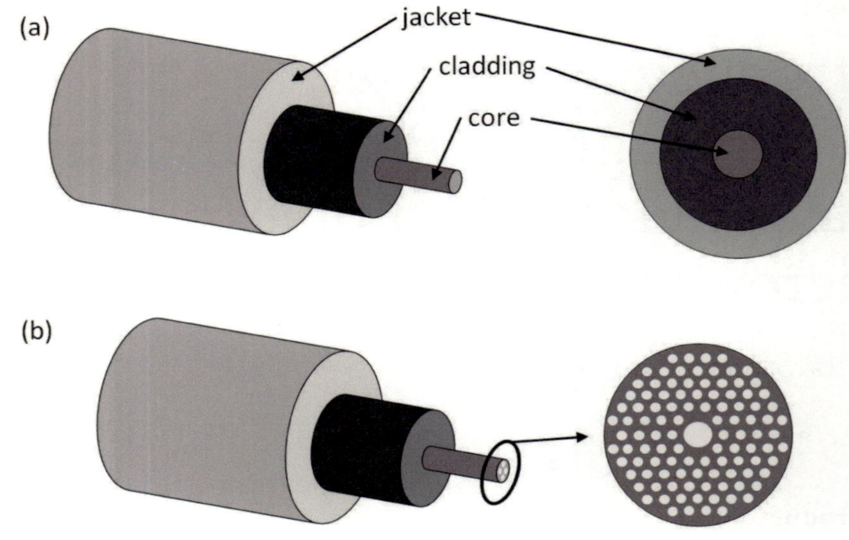

Figure 1.
Optical fiber with simple homogeneous fiber core (a) and with photonic crystals, also known as hollow core fiber (b).

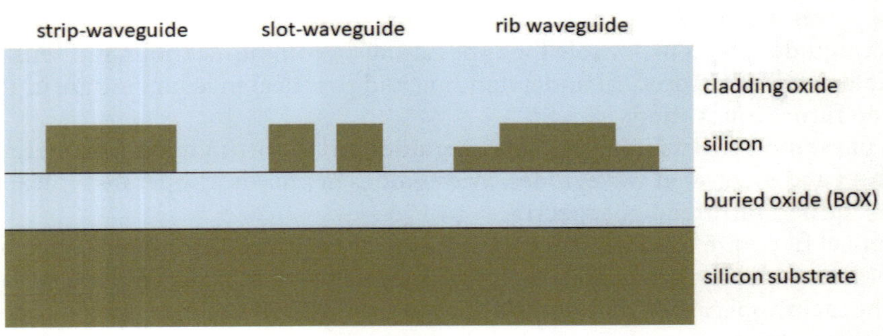

Figure 2.
Integrated optical waveguides based on silicon used in photonic integrated circuits.

modulation bandwidth [31]. This technology mainly uses slot waveguides because they provide a large overlap of optical and electrical field. Novel waveguide structures like slot waveguides [32–35] allow also the use of the quadratic electro-optical effect [36–38] and the electric field-induced linear electro-optical effect [39–40]. This gives perspective to novel modulation schemes and applications in programmable photonics.

Before optical waveguides were integrated into semiconductor chips, metal lines were already implemented several years ago, forming microwave waveguides [41]. These waveguides are used in microwave engineering. The short wavelengths distinguish microwave engineering from electronics. One particular example of microwave waveguides is the so called hollow metal pipe. A hollow metal pipe is a waveguide for electromagnetic waves, typically in the frequency range from 1 to 200 GHz [42]. Such waveguides are metal tubes with a generally rectangular, circular, or elliptical cross section. They have been studied and applied to industrial applications since almost one century [43]. New fabrication methods like 3D printing led to a renewed attention on this type of waveguide [44]. For example, practical work on microwaves concentrated on the low frequency end of the radio spectrum because it allows a long-range communication [45].

Author details

Patrick Steglich[1,2]

1 IHP—Leibniz-Institut für Innovative Mikroelektronik, Frankfurt (Oder), Germany

2 Technische Hochschule Wildau, Wildau,

*Address all correspondence to: patrick.steglich@th-wildau.de

References

[1] Steglich P, De Matteis F. Introductory chapter: Fiber optics. In: Fiber Optics-from Fundamentals to Industrial Applications. Rijeka: IntechOpen; 2019

[2] Steglich P, Heise K. Photonik einfach erklärt: Wie Licht die Industrie revolutioniert. Heidelberg: Springer-Verlag; 2019

[3] Duval S et al. Femtosecond fiber lasers reach the mid-infrared. Optica. 2015;**2**(7):623-626

[4] Zhou J et al. Intensity modulated refractive index sensor based on optical fiber Michelson interferometer. Sensors and Actuators B: Chemical. 2015;**208**:315-319

[5] Lee, Ha B, et al. Interferometric fiber optic sensors. Sensors. 2012;**12.3**:2467-2486

[6] Firstov SV et al. A 23-dB bismuth-doped optical fiber amplifier for a 1700-nm band. Scientific Reports. 2016;**6**:28939

[7] Hernaez M, Zamarreño CR, Melendi-Espina S, Bird LR, Mayes AG, Arregui FJ. Optical fibre sensors using graphene-based materials: A review. Sensors. 2017;**17**:155

[8] Zheng Y et al. Optical fiber magnetic field sensor based on magnetic fluid and microfiber mode interferometer. Optics Communications. 2015;**336**:5-8

[9] Gao R et al. Humidity sensor based on power leakage at resonance wavelengths of a hollow core fiber coated with reduced graphene oxide. Sensors and Actuators B: Chemical. 2016;**222**:618-624

[10] Hernández-Romano I et al. Optical fiber temperature sensor based on a microcavity with polymer overlay. Optics Express. 2016;**24**(5):5654-5661

[11] Ricciardi A et al. Lab-on-fiber technology: A new vision for chemical and biological sensing. Analyst. 2015;**140**(24):8068-8079

[12] Liu X, Zhang Y. Optical fiber probe-based manipulation of cells. In: Fiber Optics - from Fundamentals to Industrial Applications, Patrick Steglich and Fabio De Matteis. Rijeka: IntechOpen; 2018

[13] Shuto Y. Cavity generation modeling of fiber fuse in single-mode optical fibers. In: Fiber Optics—From Fundamentals to Industrial Applications, Patrick Steglich and Fabio De Matteis. Rijeka: IntechOpen; 2018

[14] Michel YP, Lucci M, Casalboni M, Steglich P, Schrader S. Mechanical characterisation of the four most used coating materials for optical fibres. In: 2015 International Conference on Photonics, Optics and Laser Technology (PHOTOPTICS). Berlin: IEEE; 2015. pp. 91-95

[15] Bahl M. Structured light fields in optical fibers. In: Fiber Optics—From Fundamentals to Industrial Applications, Patrick Steglich and Fabio De Matteis. Rijeka: IntechOpen; 2019

[16] Pulwer S, Fiebelkorn R, Zesch C, Steglich P, Villringer C, Villasmunta F, et al. Endoscopic orientation by multimodal data fusion. In: Proceeding SPIE 10931, MOEMS and Miniaturized Systems XVIII, 1093114. 4 March 2019. DOI: 10.1117/12.2508470

[17] Pulwer s, Steglich P, Villringer C, Bauer J, Burger M, Franz M, et al. Triangulation-based 3D surveying borescope. In: Proceeding SPIE 9890, Optical Micro- and Nanometrology VI, 989009. 26 April 2016. DOI: 10.1117/12.2225203

[18] Mai A, Steglich P, Mai C, Simon S, Scholz R. Electronic-photonic

wafer-level technologies for fast prototyping and application specific solutions. In: 2019 PhotonIcs & Electromagnetics Research Symposium—Spring (PIERS-Spring). Rome, Italy: IEEE; 2019. pp. 249-255

[19] Alimonti G et al. Use of silicon photonics wavelength multiplexing techniques for fast parallel readout in high energy physics. Nuclear Instruments and Methods in Physics Research Section A: Accelerators, Spectrometers, Detectors and Associated Equipment. 2019;**936**:601-603

[20] Steglich P et al. Hybrid-waveguide ring resonator for biochemical sensing. IEEE Sensors Journal. 2017;**17**(15):4781-4790

[21] Mai A, Bondarenko S, Mai C, Steglich P. Photonic thermal sensor integration towards electronic-photonic-IC technologies. In: ESSDERC 2019—49th European Solid-State Device Research Conference (ESSDERC). Cracow, Poland: IEEE; 2019. pp. 254-257

[22] Steglich P, Villringer C, Pulwer S, Casalboni M, Schrader S. Design optimization of silicon-on-insulator slot-waveguides for electro-optical modulators and biosensors. In: Ribeiro P, Raposo M, editors. Photoptics 2015. Springer Proceedings in Physics. Vol. 181. Cham: Springer; 2016

[23] Steglich P, Hülsemann M, Dietzel B, Mai A. Optical biosensors based on silicon-on-insulator ring resonators: A review. Molecules. 2019;*24*:519

[24] Steglich P, Mai C, Mai A. Silicon-organic hybrid photonic devices in a photonic integrated circuit technology. ECS Journal of Solid State Science and Technology. 2019;**8**(11):Q217

[25] Mai C, Steglich P, Mai A. Adjustment of the BEOL for back side module integration on wafer level in a silicon photonic technology. In: MikroSystemTechnik 2019. Berlin, Germany: Congress; 2019. pp. 1-4

[26] Steglich P et al. (keynote) silicon-organic hybrid photonics: Integration of electro-optical polymers in a photonic integrated circuit technology. ECS Transactions. 2019;**92.4**:187

[27] Steglich P, Mai C, Stolarek D, Lischke S, Kupijai S, Villringer C, et al. Partially slotted silicon ring resonator covered with electro-optical polymer. In: Proceeding SPIE 9891, Silicon Photonics and Photonic Integrated Circuits V, 98910R. 13 May 2016. DOI: 10.1117/12.2217725

[28] Steglich P et al. Functionalized materials for integrated photonics: Hybrid integration of organic materials in silicon-based photonic integrated circuits for advanced optical modulators and light-sources. In: 2019 PhotonIcs & Electromagnetics Research Symposium—Spring (PIERS-Spring). Rome, Italy: IEEE; 2019. pp. 3019-3027

[29] Steglich P et al. Novel ring resonator combining strong field confinement with high optical quality factor. IEEE Photonics Technology Letters. 2015;**27**(20):2197-2200

[30] Koeber S et al. Femtojoule electro-optic modulation using a silicon–organic hybrid device. Light: Science & Applications. 2015;**4**(2):e255-e255

[31] Alloatti L et al. 100 GHz silicon–organic hybrid modulator. Light: Science & Applications. 2014;**3**(5):e173-e173

[32] Steglich P, You KY. Silicon-on-insulator slot waveguides: Theory and applications in electro-optics and optical sensing. In: Emerging Waveguide Technology. Rijeka: IntechOpen; 2018. pp. 187-210

[33] Steglich P, Villringer C, Dümecke S, Michel YP, Casalboni M, Schrader S. Silicon-on-insulator slot-waveguide

design trade-offs. In: 2015 International Conference on Photonics, Optics and Laser Technology (PHOTOPTICS). Berlin: IEEE; 2015. pp. 47-52

[34] Bondarenko S, Villringer C, Steglich P. Comparative study of nano-slot silicon waveguides covered by dye doped and undoped polymer cladding. Applied Sciences. 2019;**9**:89

[35] Bondarenko S, Steglich P, Schrader S, Mai A. Simulation study of released silicon-on-insulator slot waveguides in a photonic integrated circuit technology. In: 2019 PhotonIcs & Electromagnetics Research Symposium—Spring (PIERS-Spring). Rome, Italy: IEEE; 2019. pp. 3334-3337. DOI: 10.1109/PIERS-Spring46901.2019.9017643

[36] Steglich P, Mai C, Villringer C, Pulwer S, Casalboni M, Schrader S, et al. Quadratic electro-optic effect in silicon-organic hybrid slot-waveguides. Optics Letters. 2018;**43**:3598-3601

[37] Steglich P et al. On-chip dispersion measurement of the quadratic electro-optic effect in nonlinear optical polymers using a photonic integrated circuit technology. IEEE Photonics Journal. June 2019;**11**(3):1-10

[38] Steglich P et al. Quadratic electro-optical silicon-organic hybrid RF modulator in a photonic integrated circuit technology. In: 2018 IEEE International Electron Devices Meeting (IEDM). San Francisco, CA: IEEE; 2018. pp. 23.3.1-23.3.4

[39] Steglich P et al. Electric field-induced linear electro-optic effect observed in silicon-organic hybrid ring resonator. IEEE Photonics Technology Letters. 2020;**32**(9):526-529

[40] Steglich P et al. Direct observation and simultaneous use of linear and quadratic electro-optical effects. Journal of Physics D: Applied Physics. 2020;**53**(12):125106

[41] Davidson DB. Computational Electromagnetics for RF and Microwave Engineering. Cambridge: Cambridge University Press; 2010

[42] Pozar DM. Microwave Engineering. 4th Ed. New Jersey: Wiley; 2011

[43] Chu LJ. Electromagnetic waves in elliptic hollow pipes of metal. Journal of Applied Physics. 1938;**9**(9):583-591

[44] D'Auria M, Otter WJ, Hazell J, Gillatt BT, Long-Collins C, Ridler NM, et al. 3-D printed metal-pipe rectangular waveguides. IEEE Transactions on Components, Packaging and Manufacturing Technology. 2015;**5**(9):1339-1349

[45] Packard KS. The origin of waveguides: A case of multiple rediscovery. IEEE Transactions on Microwave Theory and Techniques. 1984;**32**(9):961-969

Section 2

Waveguides in Photonics

Chapter 2

The Antiresonant Reflecting Optical Waveguide Fiber Sensor

Ran Gao and Jiansen Ye

Abstract

In this chapter, the optical fiber sensors based on antiresonant reflecting optical waveguide have been introduced, including the single layer, double layers, double resonators, and hybrid mechanism. Various optical fiber sensors based on antiresonant reflecting optical waveguide have been introduced in this chapter with different working principles, including the fiber optic vibration sensor, humidity sensor, strain sensor, temperature sensor, magnetic field sensor, biosensor, etc. Especially, many long-standing challenges in the fiber optic sensor can be solved through the working principle of the antiresonant reflecting optical waveguide, including the temperature cross-talk compensation, distribution localization, optofluidic biosensing, etc. In general, the optical fiber sensors based on antiresonant reflecting optical waveguide have advantages, such as compact structure, high sensitivity, large dynamic range, and high stability, which appear to have potential applications in researches of structure health monitoring, oil exploiting, and biology detection.

Keywords: antiresonant reflecting optical waveguide, Fabry-Pérot resonator, double layers, double resonators, hybrid mechanism

1. Background

Over past two decades, the antiresonant reflecting optical waveguide (ARROW) has developed into a versatile platform for a range of interdisciplinary applications in low loss communication [1], ultrafast optics [2], optical amplifiers [3], and biophotonics [4]. In the ARROW, the guided light is reflected at the two surfaces of the cladding in the hollow-core fiber (HCF), forming a Fabry-Pérot etalon [5]. The guided light at the antiresonant wavelength can be propagated along the HCF. Due to the unique light guiding mechanism, the ARROW is a good candidate for the fiber optic sensor: (i) the optical properties of the ARROW can be easily manipulated with the cladding structure, making the flexibility for fiber optic sensors; (ii) the guided light can break the confining of the fiber core, forming an enhanced interaction between the light and the ambient medium; and (iii) the hollow holes in the HCF is a natural channel for the optofluidic biosensors, which reduce the complexity of the fiber optic sensor significantly. Many sensing principles of ARROWs for fiber optic sensors have been researched in recent years, including the ARROW with the single layer, double layers, double resonators, hybrid mechanism, etc. [6]. The ARROW-based fiber optic sensors possess great flexibility, high sensitivity, and low cost, which are expected to be used for many fields in real-world

Sensors	Performance	Application
Vibration sensor	(1) Signal-to-noise ratio of 60 dB, (2) wide frequency response from 5 to 10 kHz, and (3) high sensitivity and low temperature cross-sensitivity	Applications for the monitoring of smart structures such as buildings, bridges, highways, pavements, dams, and so on
Humidity sensor	(1) The sensitivity of up to 0.22 dB/% RH and (2) good repeatability, fast response time, and low temperature cross-sensitivity	Applications for human daily biology, industrial production, agriculture, animal husbandry, and other fields
Water-level sensor	Sensitivity of 1.1 dB/mm	Applications for water-level monitoring
Strain sensor	(1) The resolution of the sensor is 27.9 pm/με and (2) the temperature cross-sensitivity is only 1.67 pm/°C	Applications for the accurate measurement of the strain of smart structures
Magnetic field sensor	(1) Magnetic field sensitivity of 81 pm/Oe and (2) low temperature cross-sensitivity	Applications for industrial production, motor, and electronic products' magnetic field measurement
Optic biosensors	(1) The limit of detection of 0.5 ng/ml can be achieved for the IFN-γ concentration and (2) the influence of the temperature could be compensated through the referenced resonance dip	Applications for health monitoring, cancer prevention, biological engineering, etc.
Pressure sensor	(1) The pressure sensitivity of −4.42 nm/MPa and (2) the spatial sensitivity of 0.86 nm/cm can be achieved	Applications for multipoint pressure detection in the fields of security, structure monitoring, and oil exploration
Temperature sensor	Temperature sensitivity of 70.71 pm/°C	Applications for biomedicine, industrial production, space exploration, and other fields

Table 1.
The optical fiber sensor based on ARROW [7–15].

applications [5, 6]. The optical fiber sensor based on the principle of ARROW is mainly applied in the optical fiber vibration sensor, optical fiber humidity sensor, water-level sensor, fiber strain sensor, fiber optic magnetic field sensor, fiber optic biosensors, optical fiber pressure sensor, optical fiber temperature sensor, and other types of optical fiber sensors. The common optical fiber sensors based on the ARROW principle and their performance are shown in **Table 1**.

2. The principle of the antiresonant reflecting optical waveguide

The antiresonance reflection principle refers to light that does not meet the resonance condition and is confined to the low refractive index (RI) fiber core for transmission [5]. In 1986, the antiresonant reflecting optical waveguide (ARROW) was proposed by Duguya et al. [5]. The working principle of the ARROW is to use the design of the multilayer high reflection film between the waveguide and silicon substrate (most of which use the double-layer film) to transmit the light beam in the waveguide layer, so as to reduce the energy leakage, with the characteristics of single mode and small loss [5, 6, 16]. The ARROW structure is a promising waveguide structure for silicon-based sensors and has been used for a variety of purposes because it allows the thickness of the buffer layer to be reduced and the single-mode size to be increased and the process tolerance and material selection range to be relatively expanded [17–19].

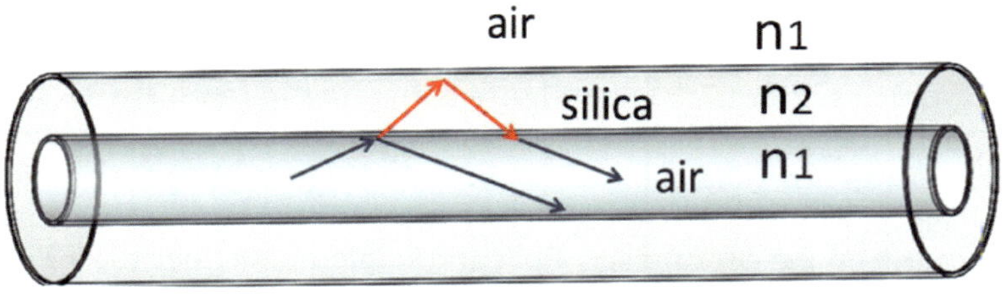

Figure 1.
The structure of ARROW.

The ARROW can be used to study the optical conduction mechanism of low-index core fibers, which is similar to Fabry-Pérot resonators [6]. **Figure 1** shows the structure diagram of the ARROW, where the gray part is the high-index layer with RI n2 (silica) corresponding to the Fabry-Pérot resonator cavity and the dark gray part is the low-index layer with RI n1 (air) [17, 18]. When the wavelength of the optical fiber satisfies the resonance condition, it will leak out from the high fold rate layer corresponding to the low transmission intensity part of the transmission spectrum, which is similar to the destructive interference of light in the Fabry-Pérot cavity [19, 20]. In contrast, the light of the antiresonance wavelength will be reflected back at the interface of the high and low RIs and will be restricted to be transmitted in the fiber core. So most of the light will be reflected back to the fiber core [20]. Based on the principle of the antiresonant waveguide, the optical fiber sensor can be made into different types by changing the relevant parameters. Optical fiber sensors based on the antiresonant waveguide are mainly divided into two types [21]. The first is to change the resonant wavelength by changing the RI or the length of the Fabry-Pérot cavity. The second is to change the contrast of the output intensity of the antiresonant waveguide by changing the associated external material composition [19, 20]. The optical fiber sensor of the ARROW finally realizes the measurement of different physical quantities through these two kinds of modulation methods. This chapter will introduce the working principles of various ARROW-based fiber optic sensors in detail [20, 21].

3. The antiresonant reflecting optical waveguide sensor

3.1 The single-layered antiresonant reflecting optical waveguide sensor

The simplest model of the antiresonant reflecting optical waveguide (ARROW) is the single-layered antiresonant reflecting optical waveguide (SL-ARROW), which is widely used in optical fiber sensors. The SL-ARROW mode in the capillary waveguide is sensitive to the surrounding environments, and various sensing applications have been proposed, such as vibration sensor, humidity sensor, water-level sensing, etc. The working principle of the SL-ARROW is introduced by the vibration sensor [7].

Vibration signal detection is very important in the application of structural monitoring in people's life. Vibration detection can monitor the safety of building structures such as buildings, highways, bridges, dams, highways, and so on. In general, the vibration can be detected by piezoelectric, magneto-electric, and current sensors [22]. SL-ARROW is an optical fiber vibration sensor, which is designed based on the principle of antiresonant waveguide. The fiber optic sensor is a good alternative with several unique advantages such as low weight, immunity to electromagnetic interference, and long-distance signal transmission for remote operation [23]. Here, an all-fiber vibration sensor based on the SL-ARROW has been proposed in a tapered capillary fiber. The schematic construction of the proposed vibration sensor is given in **Figure 2**. In order to fabricate the sensor, the capillary optical fiber is selected as the sensing fiber. The capillary fiber consists of a hollow core with an inner diameter of 30 μm and a ring-cladding with a thickness of 55 μm. Both ends of the 8 cm long capillary optical fiber are cut by the high-precision cutter, and the cut capillary fiber and single-mode fiber (SMF) are spliced through the fiber splicer. The cross section of the capillary fiber and the splicing diagram between the capillary fiber and SMF are shown in **Figure 2** [7].

The principle of using capillary fiber to make vibration sensor can be described as an SL-ARROW [24]. As the RI of the cladding is larger than that of the core, the core mode can oscillate and radiate through the cladding. The cladding modes propagate in the cladding region of the capillary fiber, as shown in **Figure 3(a)**. The working principle of the tapered capillary fiber can be approximated to Fabry-Pérot etalon, as shown in **Figure 3(b)**. When the wavelengths cannot satisfy the resonant condition, the optical waveguide will be confined in the hollow core of the fiber as the core modes. Therefore, the guide light can be reflected by the resonator. On the contrary, when the wavelength meets the resonant condition, the guided light will resonate and will not be reflected by the Fabry-Pérot cavity and will leak out of the

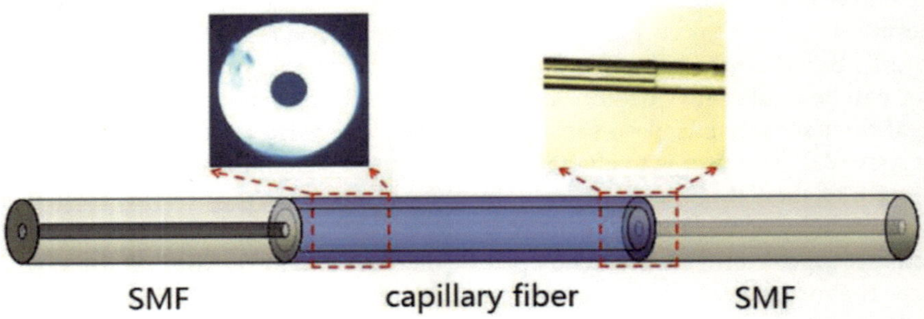

Figure 2.
The schematic construction of the proposed sensor [7].

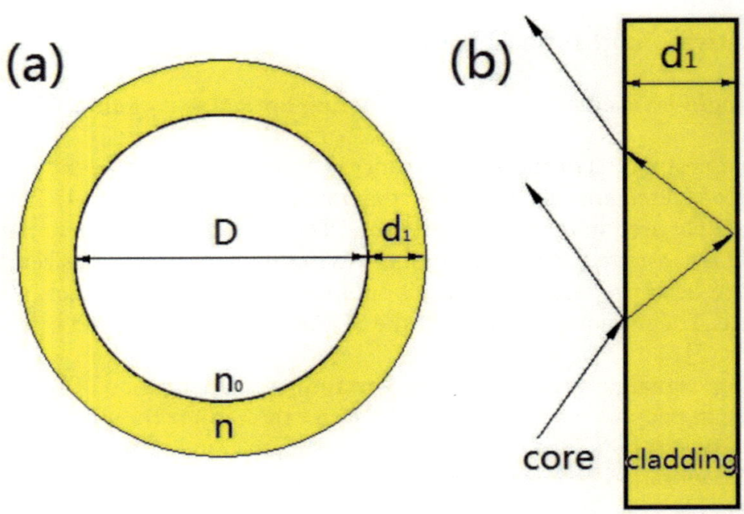

Figure 3.
(a) Schematic diagram of the cross section of the capillary fiber and (b) the guiding mechanism of the capillary fiber [7].

cladding to generate the transmission spectrum. In the transmission spectrum, periodic and narrow loss attenuation corresponding to the Fabry-Pérot cavity resonance condition will appear [7]. The wavelength of the lossy dip corresponding to the resonance condition λ_r can be expressed as Eq. (1) [7, 22]:

$$\lambda_r = \frac{2d\sqrt{n^2 - n_0^2}}{m} \tag{1}$$

where n and n_0 are the RIs of the capillary fiber cladding and air, respectively, d is the thickness of capillary fiber cladding, and m is the resonance order [7].

The mechanical analysis of the photoelastic effect of the tapered region under bending can be expressed as follows [7]:

$$\Delta n = \frac{1}{4} n^2 (1 + \gamma)(P_{12} - P_{11}) \left(\frac{A}{R}\right)^2 \tag{2}$$

where Δn is the change of the RI, n is the RI of the silica, P_{11} and P_{12} are the optoelastic constants of silica, γ is the Poisson ratio of silica, A is the diameter of the tapered region, and R is the bending radius. Substitution of Eq. (1) into Eq. (2) gives

$$\lambda_r = \frac{2(d\sqrt{\left(n + \frac{1}{4}n^2(1+\gamma)(P_{12} - P_{11})\left(\frac{A}{R}\right)^2\right)^2 - n_0^2}}{m} \tag{3}$$

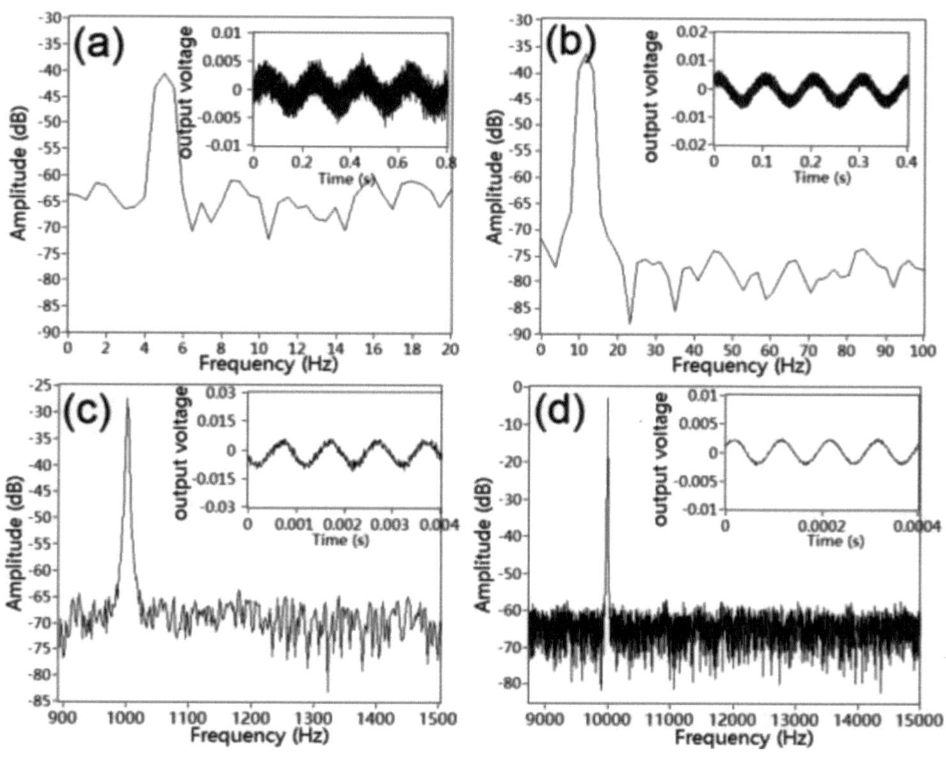

Figure 4.
The frequency spectrum of the tapered capillary fiber corresponding to (a) 5 Hz, (b) 10 Hz, (c) 1 kHz, and (d) 10 kHz [25].

According to Eq. (3), it can be seen that the wavelength of the loss dip in the transmission spectrum under resonance condition is highly sensitive to the small change of bending radius in the tapered region. This provides an attractive method for vibration detection.

To test the frequency response of the fiber vibration sensor, the corresponding frequency response experiments were performed, and the transmission spectra are shown in **Figure 4** [7]. In the fiber vibration sensor experiment, the piezoelectric plate was used as the vibration source, which was driven by 1.8 V sinusoidal signals of 5.0 Hz, 10.0 Hz, 1.0 kHz, and 10.0 kHz, respectively. The transmission intensity of the fiber sensor was converted into a frequency spectrum by fast Fourier transform (FFT), as shown in **Figure 4(a)–(d)** [7]. The corresponding time domain signals are shown in the insets. The sampling rate of each frequency is 1 M and the total sampling time is 5 s. All of transmission intensities of sensor were modulated to sinusoidal waveforms with very uniform amplitude [7]. The main peaks of the frequency spectrum are located at 4.98, 9.98, 998.03, and 9998.96 Hz, respectively, which are close to the corresponding driving frequency [25].

3.2 The double-layered antiresonant reflecting optical waveguide

Compared with the SL-ARROW, the Fabry-Pérot resonator in the ARROW can be also formed through two layers with different claddings of the fiber. A double-layered Fabry-Pérot resonator can be formed between the silica cladding and polymethyl methacrylate (PMMA) cladding [9, 10]. The DL-ARROW possesses high flexibility, which can be used in various sensing applications such as strain sensor, film sensing, temperature, vibration sensing, etc. Taking the temperature sensor as an example, the working principle of the double-layered ARROW (DL-ARROW) is introduced [9].

In real life, optical fiber temperature sensor has been widely used in temperature measurement in different application areas. Optical fiber temperature sensor has many unique advantages such as immunity to electromagnetic interferences, high sensitivity, repeatability, stability, durability, high resolution and fast response, durability against harsh environments, and other advantages [26]. A temperature sensor based on the DL-ARROW was presented [9]. The cross-sectional view of the fiber is given in **Figure 5 (a)**. A section of the simplified hollow-core (SHC) photonic crystal fiber (PCF) was cleaved at both ends, and one end of the fiber was sealed with glue, as can be seen in **Figure 5(b)**, the selectively opened air hole can be seen in **Figure 5(c)**, and the schematic construction of the proposed device is given in **Figure 5(d)** [9].

The essence of the optical fiber temperature sensor guidance is mainly driven by the silicon rod around the hollow core, which plays the role of the ARROW [25]. Before the fiber is penetrated by alcohol, due to the weak interaction between the core and the cladding mode, light is well restricted in the air core of the SHC-PCF. This can be interpreted as a strong lateral field mismatch between the modes, resulting in the overlap with the core field distribution being washed away. **Figure 6** gives the sketch of the optical path of the beams at the alcohol-filled area and the outer silica cladding [9].

In the presence of alcohol in the SHC-PCF cavity, due to the better phase matching, the interaction between the core layer and the cladding mode can be greatly enhanced, resulting in the degradation of the optical field constraints in the hollow core. The core mode field will radiate through the silicone ring around the air core to the outer cladding. The alcohol-filled cavity combined with the external silicon cladding can be regarded as a double-layer Fabry-Pérot resonant cavity

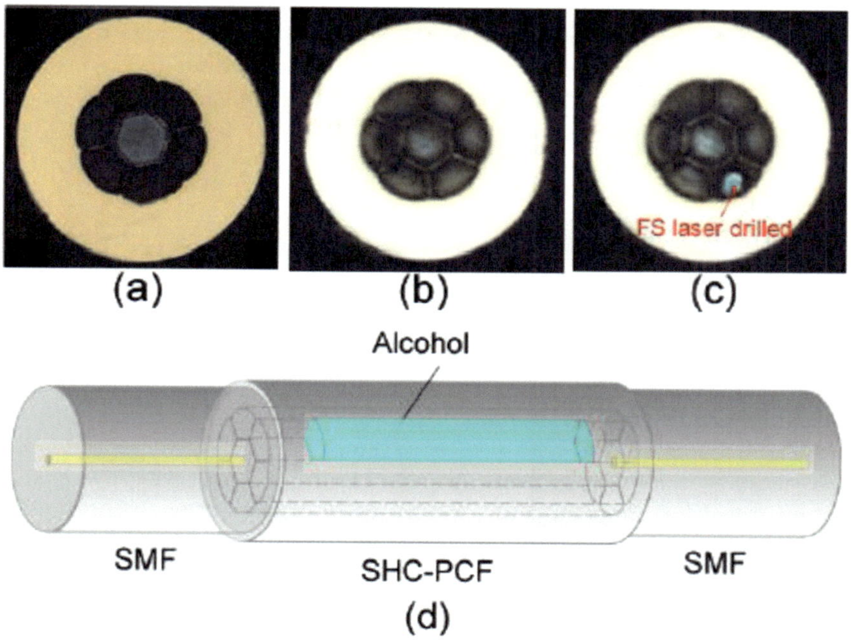

Figure 5.
(a) The cross-sectional view of the SHC-PCF, (b) the SHC-PCF sealed at the end face, (c) with end face sealed selectively opened, and (d) the structure diagram of the sensor.

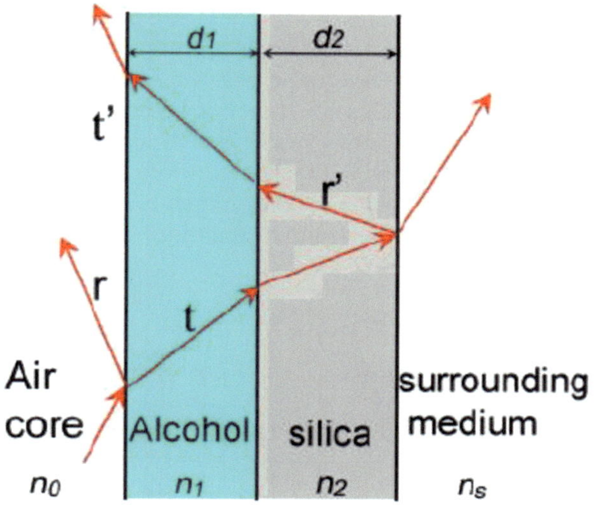

Figure 6.
The optical path of the interference beams at the alcohol-filled area and the outer cladding [10].

[9, 27]. For wavelengths that satisfy the resonance conditions of the resonator, constructive interference occurs, which means that the Fabry-Pérot resonator is highly transparent to these wavelengths, and light cannot be reflected and will leak out of the cladding [27]. In contrast, for the antiresonance wavelength (i.e., the

wavelength that does not satisfy the resonance condition), destructive interference occurs, and light can be well reflected by the Fabry-Pérot resonator [9]. The Fabry-Pérot resonator is confined within the cavity of the optical fiber and serves as a guide for the waveguide mold. The position of the non-transmitted wavelength can be described by the following formula as Eq. (4) [9, 27]:

$$\lambda_m = \frac{2(d_1\sqrt{n_1^2 - n_0^2} + d_2\sqrt{n^2{}_2 - n_0^2}}{m} \tag{4}$$

where d_1 is the thickness of the air cladding, d_2 is the thickness of the outer cladding, and m is a positive integer. n_0, n_1 and n_2 are the refractive indices of air, alcohol and silica, respectively. From Eq. (4), it can be seen that the resonant wavelengths are mainly determined by the thickness of the alcohol and the outer silicon cladding and the material index.

By differentiating Eq. (4) with respect to temperature (T), we get the temperature sensitivity as in Eq. (5). It should be noted that in Eq. (5), we only considered the change of n_1 ($dn_1 = \alpha dT$, α is the thermo-optic coefficient of alcohol), since silica has a much smaller thermo-optic coefficient than alcohol, which can be ignored in this case as Eq. (5) [9]:

$$\frac{\partial \lambda_m}{\partial T} = \frac{2d_1 n_1}{m\sqrt{n_1^2 - n_0^2}}\alpha \tag{5}$$

The thermo-optic coefficient of alcohol is -3.99×10^{-4}.

The temperature response characteristics of this temperature sensor were investigated in a temperature box with adjustable temperature. The temperature sensor with a filling length of 1.2 cm is put into a temperature box. The temperature of the temperature box was heated from room temperature to 60° with an increment of 10°C. The maximum value of temperature measurement is limited below the boiling point of alcohol (78.37°) [9]. This situation can be improved by using a liquid with higher boiling point like ethylene glycol [9]. The attenuation dips were found to shift toward shorter wavelengths with temperature increasing, as can be seen in **Figure 7(a)** [9]. The wavelength shift versus temperature in the experiment is plotted in **Figure 7(b)**, and the temperature sensitivity was obtained to be −0.48 nm/°. [9].

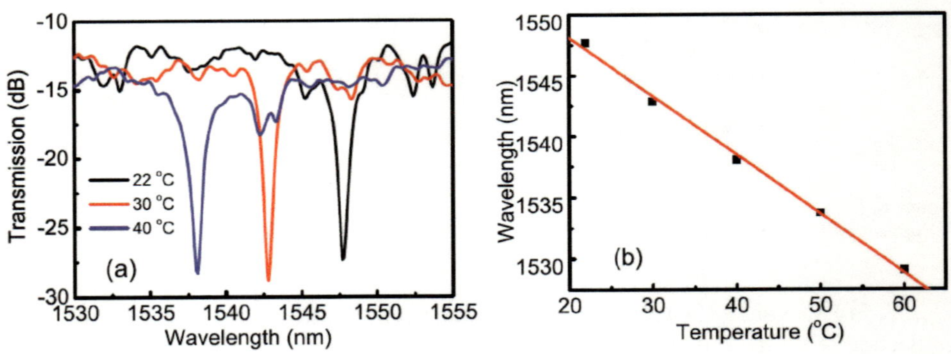

Figure 7.
(a) Spectrum blueshift with increasing of temperature and (b) temperature response of the SHC-PCF with filling length of 2 cm [9].

3.3 The double resonator antiresonant reflecting optical waveguide sensor

Both the SL-ARROW and DL-ARROW are formed to generate the ARROW effect in the fiber optic sensor. However, many long-standing challenges still exist in the ARROW-based fiber sensor, such as serious temperature cross-sensitivity [12]. The ambient temperature change can also modulate the resonance condition of ARROW, making a decrease of the measurement accuracy for both SL-ARROW and DL-ARROW. Therefore, the double resonator ARROW (DR-ARROW) has been investigated to solve the temperature cross talk in the fiber sensor. In this section, the biosensor fiber sensors are taken as examples, and the working principle of the DR-ARROW is introduced [12].

The in-line fiber optofluidic waveguide biosensors possess both enhanced sensing performance and ultracompact size, which have been widely used in the lab-in-fibers chemical and biological sensing [12]. In-line fiber biosensors possess many distinctive advantages such as high sensitivity, compact size, and immunity to electromagnetic interference [28]. Here, a biosensor based on DR-ARROW for the detection of interferon-gamma (IFN-γ) concentration has been introduced. The schematic construction of the proposed biosensor sensor is given in **Figure 8**. In the proposed sensor, a HCF was employed as the sensing fiber, as shown in **Figure 8(a)**. The HCF consists of an air octagon core, an air-ring cladding with eight holes, and a silica cladding. The NaCl solution was filled into a cladding hole with a length of 10 cm through the capillary force by immersing the remaining SMF into the NaCl

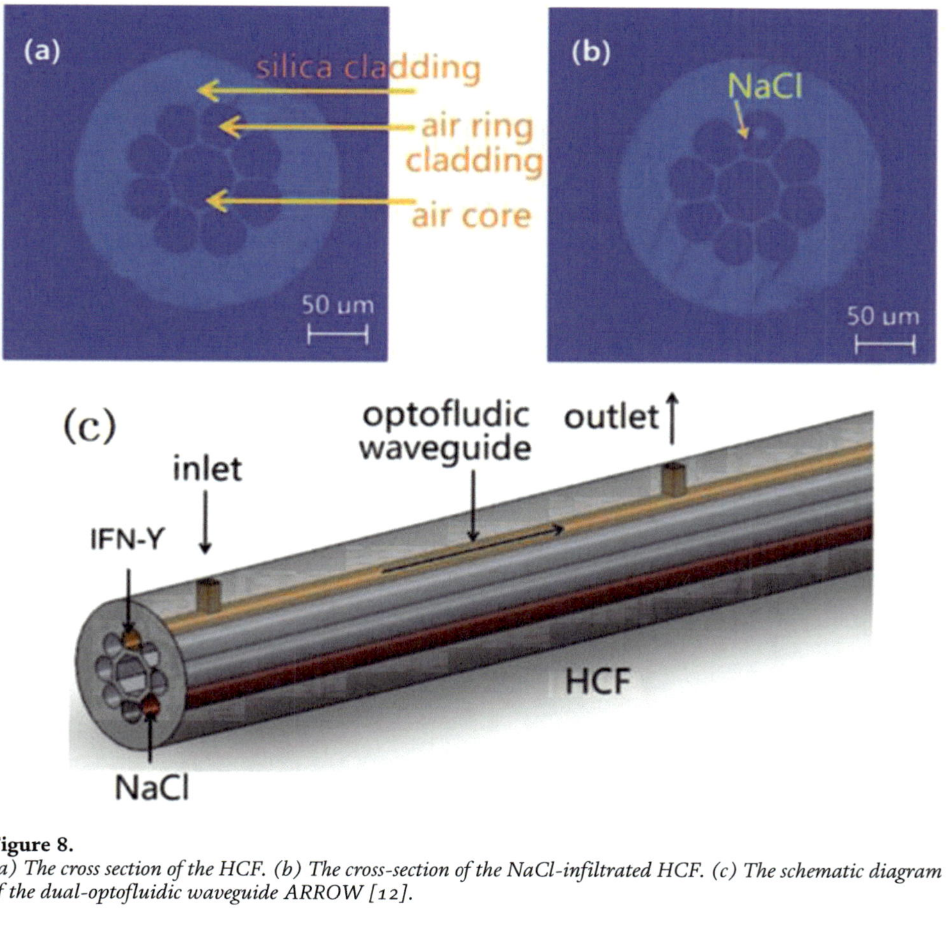

Figure 8.
(a) The cross section of the HCF. (b) The cross-section of the NaCl-infiltrated HCF. (c) The schematic diagram of the dual-optofluidic waveguide ARROW [12].

solution for ∼48 h, as shown in **Figure 8(b)**. The liquid sample could be also pumped out of the hole through the other microchannel, as an outlet, as shown in **Figure 8(c)** [12].

The cross-sectional schematic diagram of the optical fiber biosensor is shown in **Figure 9(a)**, which can be described by a DL-ARROW model [12]. Due to the infiltration of IFN-γ or NaCl with high RI in the cladding of HCF, the guided light will reflect at the two interfaces of the cladding, thus forming a Fabry-Pérot resonator, as shown in **Figure 9(b)**. Therefore, an ARROW is formed in HCF. However, due to two different infiltration materials, the IFN-γ solution and the NaCl solution, two ARROWs appear in HCF. At the wavelength of 1550.38 nm, the resonance condition of the ARROW for the optofluidic waveguide is achieved [12]. On the other hand, at the wavelength of 1557.86 nm, the resonance condition of the ARROW for the NaCl-infiltrated channel is achieved [13]. Hence, there are two resonance dips corresponding to two materials, which can be expressed as Eqs. (6) and (7) [12, 13]:

$$\lambda_o = \frac{2\left(d_{op}\sqrt{n_{op}^2 - n_{air}^2} + d_{cl}\sqrt{n_{silica}^2 - n_{air}^2}\right)}{m} \tag{6}$$

$$\lambda_n = \frac{2\left(d_{op}\sqrt{n_{na}^2 - n_{air}^2} + d_{cl}\sqrt{n_{silica}^2 - n_{air}^2}\right)}{m} \tag{7}$$

where m is the resonance order, λ_o and λ_n are wavelengths of resonance dips for IFN-γ and NaCl solution, d_{op} and d_{cl} are diameters of the hole and thickness of the silica cladding, and n_{air}, n_{op}, n_{na} and n_{silica} are RIs of the core, IFN-γ solution, NaCl solution, and silica, respectively.

Due to the change of RI in the optofluidic waveguide [12], the immunoreaction between the aptamer and the IFN-γ could modulate the wavelength of the resonant dip for the IFN-γ infiltrated ARROW. However, the resonant wavelength dip is fixed because of the channel independence of NaCl-infiltrated resonator. On the other side, due to the similar thermos-optical coefficients of the NaCl and IFN-r, the wavelength shifts of two resonance dips corresponding to two ARROWs have the same response to the temperature fluctuation [12]. The response of the two resonance attenuation wavelength shifts corresponding to the two ARROWs to temperature fluctuation is the same. Therefore, a dual-optofluidic waveguide ARROW

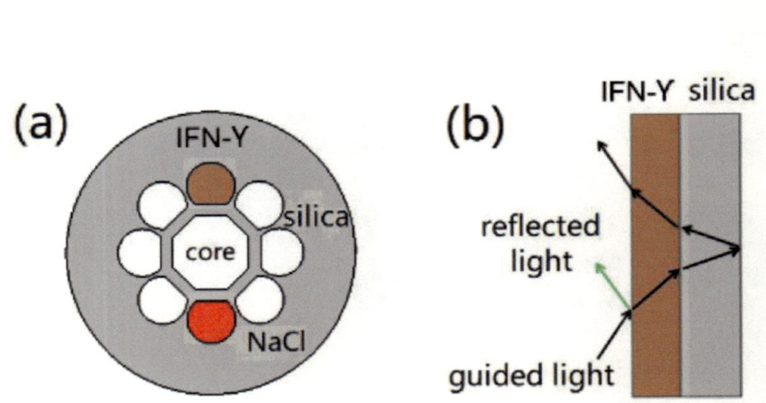

Figure 9.
(a) Diagram of the dual-optofluidic waveguide ARROW and (b) the Fabry-Pérot resonator [12].

biosensor with temperature compensation is formed by interrogating and inquiring the wavelength interval between the two resonance dips of the dual-optofluidic flow waveguide ARROW [12, 13].

The biosensor of DR-ARROW was also tested [12]. The corresponding transmission spectra, **Figure 1A** is shown in Appendix [12]. The transmission spectrum of the ethanol and NaCl infiltrated the dual-optofluidic waveguide ARROW biosensor, **Figure 1A(a)** is shown in Appendix [12]. There are two resonance dips in the wavelength range of 1525–1565 nm. The resonance dips at 1545.58 and 1550.86 nm are consistent with the theoretical predictions of the ethanol-infiltrated Fabry-Pérot cavity (1545.78 nm) and the NaCl-infiltrated Fabry-Pérot cavity (1550.38 nm) [12]. Therefore, there will be two resonance dips if the two materials infiltrated Fabry-Pérot resonators at the same time. Different concentrations of ethanol solution are injected into the optofluidic waveguide channel, and the corresponding RI changes from 1.3568 to 1.3622 RIU. When the temperature is 20° C, the wavelength shift of the ARROW of dual-optofluidic waveguide with different RIs, **Figure 1A(b)** is shown in Appendix [12]. The wavelengths of two resonance attenuations for different RIs, **Figure 1A(C)** is shown in Appendix [12]. In addition, the wavelength spacing of the two resonance dips, **Figure 1A(d)** is shown in Appendix [12]. In this experiment, if the resolution of OSA is 0.02 nm, the sensitivity of RI response can reach −1413 nm/RIU [12].

In addition, the effect of temperature on the biosensor was investigated by experiments [12]. The transmission spectra of the biosensor at different temperatures from 20 to 80°C are shown in **Figure 10(a)** [12]. At different temperatures, the two resonance decays shift to longer wavelengths at the same time. However, due to the same temperature response of the two ARROWs, the wavelength interval is maintained at a standard change of 0.02 nm, as shown in **Figure 10(b)** [12]. Hence, the dual-optofluidic waveguide ARROW biosensor is not sensitive to temperature by interrogating the wavelength interval [12].

3.4 The hybrid antiresonant reflecting optical waveguide sensor

Most of the ARROW-based fiber sensors only rely on the working principle of the ARROW. In recent years, the fiber optic sensor of the hybrid antiresonant reflecting optical waveguide (H-ARROW) has been researched. Besides the ARROW, another mechanism was also formed in the fiber, making a hybrid

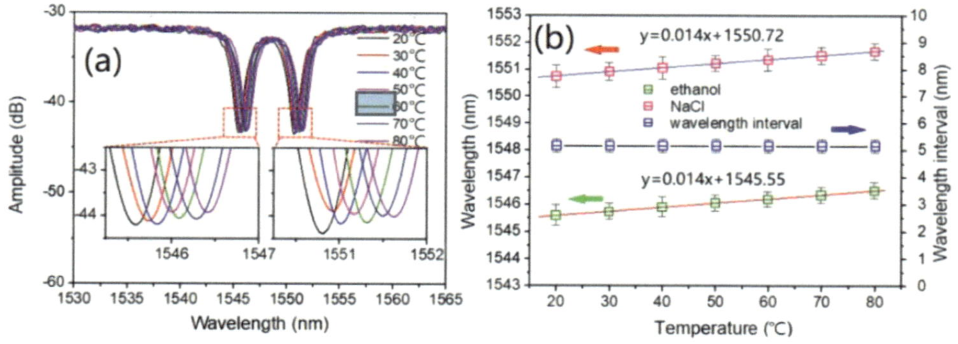

Figure 10.
(a) Wavelength shifts at different temperature from 20 to 80°C and (b) wavelength of resonance dips and the wavelength interval [12].

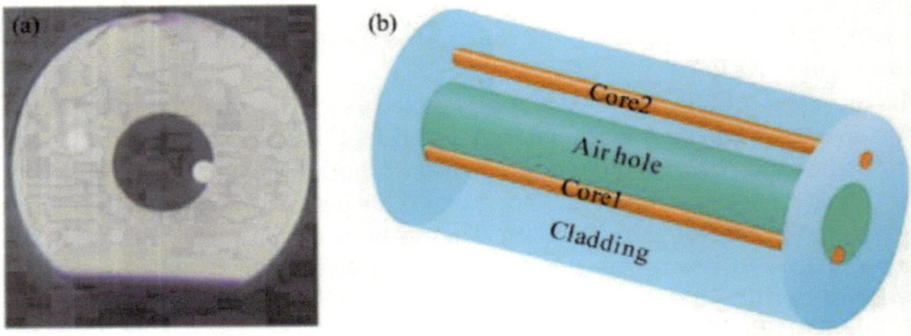

Figure 11.
(a) The cross section of single-hole twin eccentric core fiber and (b) three-dimensional diagram of single-hole twin eccentric core fiber [15].

mechanism in a single ARROW-based fiber sensor [14, 15]. The two different mechanisms could measure different parameters dependently, which increase the multifunction of the ARROW-based fiber sensor significantly. In this section, a curvature ARROW-based fiber sensor with hybrid mechanism is introduced as an example [15].

Optical fiber curvature sensors have been widely used in structural health monitoring and distributed sensing fields, such as buildings, towers, and bridges [15, 29]. The optical fiber curvature sensor has the advantages of small volume, high sensitivity, and no electromagnetic interference [30, 31]. Here, an all-fiber vibration sensor is based on H-ARROW, through the integrated antiresonance mechanism and in-line Mach Zehnder interference (MZI) [32]. The schematic construction of the proposed curvature sensor is given in **Figure 11** [15]. The single-hole double-core fiber is characterized by replacing the core with a large air hole, one of which is suspended on the inner surface of the cladding. As shown in **Figure 11(a)**, the other core is asymmetrically distributed outside the air hole. As can be seen from **Figure 11(b)**, for the curvature fiber, Core 1 is suspended in the air hole in the middle of the hollow-core fiber, and Core 2 is in the cladding of the hollow-core fiber [15].

The principle of curvature sensor of H-ARROW is analyzed [15]. According to its structure, the single-hole dual-core fiber is characterized by replacing the core with a large air hole, one of which is suspended on the inner surface of the cladding, as shown in **Figure 12** [15]. According to the working principle of the single-hole double-eccentric core fiber, the distribution of light field in the whole fiber is of great significance [15]. A section of a 2.6 mm single-hole double-eccentric core fiber is fused between two single-mode fibers. It is well known that when a beam travels between different media, it will reflect and refract. Specifically, for the beams propagating from the optical dense medium to the optical thin film medium, only when the incident angle is less than the critical incident angle, the two effects exist at the same time [15].

The transmission spectrum of the optical fiber sensor is controlled by two effects: antiresonance and in-line MZI [15]. It should be noted that the backscattered beam undergoes multiple reflections in the cladding. As a result, Fabry-Pérot resonators are formed in a silicon cladding. The antiresonance effect of the single-hole double eccentric core fiber can be regarded as the reflection type Fabry-Pérot interferometer. Therefore, the transmission of the antiresonance effect can be expressed as Eq. (8) [15, 32]:

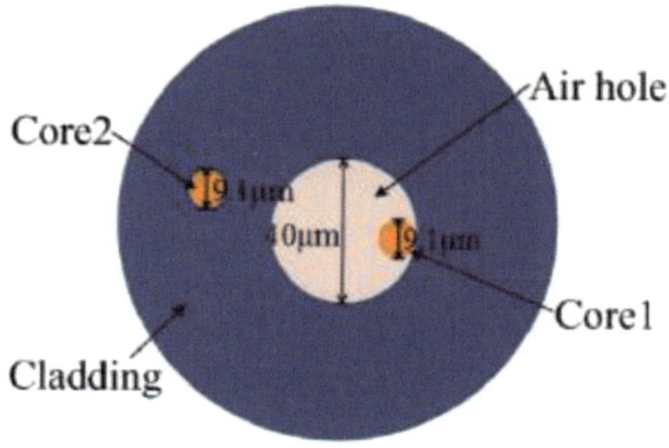

Figure 12.
The cross-sectional microscope images of the single-hole twin eccentric core fiber [15].

$$T_{anti} = A\frac{F\sin^2\left(\frac{2\pi}{\lambda}n(\lambda)l\right)}{1+F\sin^2\left(\frac{2\pi}{\lambda}n(\lambda)l\right)} \tag{8}$$

where F represents the fringe finesse coefficient of the multiple-beam interferometer. λ and $n(\lambda)$ are the wavelengths of the spectrum and effective RI of the cladding, respectively. A and l are the intensity coefficients of the whole antiresonant effect and optical path of the antiresonant beams. In addition, the wavelength at resonance can be obtained by the following equation as Eq. (9) [15, 32]:

$$\lambda_{anti} = \frac{2d}{m}\sqrt{n_1^2 - n_2^2} \tag{9}$$

where d is the thickness of the single-hole twin eccentric core fiber cladding, and m is the resonance order. n_1 and n_2 are the RIs of the single-hole twin eccentric core fiber cladding and the air, respectively.

The sensor structure also produces MZ interference rather than the antiresonance effect. As shown in **Figures 11(a)** and **(b)**, the in-line MZ structure forms several modes, including core mode, low-order mode, and high-order cladding mode. The dominant interference is formed between the core mode and the low-order cladding mode. Therefore, the transmission of the multimode interference can be normalized as Eq. (10) [15, 32]:

$$T_{MZi} = B_i\cos^2\left(\frac{\pi}{\lambda}\cdot\Delta n_i\cdot L\right) \tag{10}$$

where B_i is the intensity coefficient of the comb spectrum, Δn_i is the effective RI difference between the fiber core mode and cladding modes, L is the physical length of the special fiber cladding, and i represents the order of the cladding modes.

A curvature experiment is conducted to investigate the sensing properties [15]. The test results are shown in **Figure 13**. The sensing characteristics are investigated by the curvature experiment [15]. The test results are shown in **Figure 13**. The curvature variation can be derived from 0.94 to 2.10 m^{-1} according to the equation of $Rsin(L/2R) = (L - d)/2$ [15]. In this experiment, fitted resonant wavelength

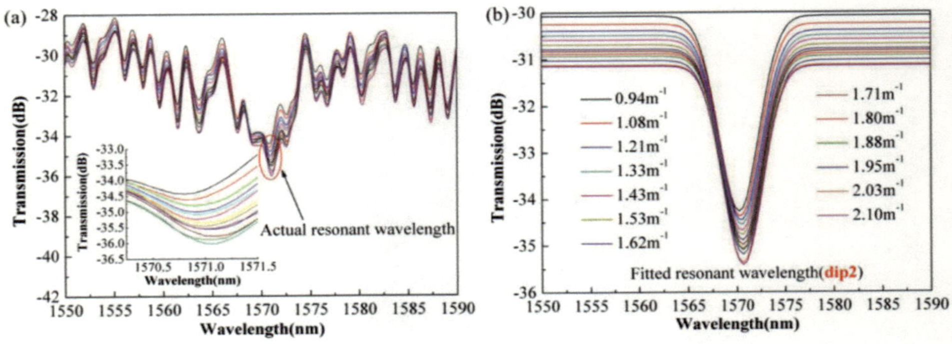

Figure 13.
(a) The intensity variation of the actual resonant wavelength with the curvature increased and (b) the intensity variation of the Gaussian fits resonant wavelength (dip2) with curvature increasing from 0.94 to 2.1 m^{-1} *[15].*

(dip2) is selected to monitor the curvature variation trend, corresponding to the resonant wavelength of 1570 nm [15]. The intensity of the actual wavelength decreases as the curvature increases, as shown in **Figure 13**. The intensity of the actual resonant wavelength decreases as the curvature increases, as shown in **Figure 13(a)** [15]. The actual wavelength circled by the red ellipse is about 1571 nm, and the inset is the enlarged view of the intensity variation [15]. The actual wavelength circled by the red ellipse is about 1571 nm, and the illustration is an enlarged picture of intensity change [15]. **Figure 13(b)** reflects that there is also only intensity variation without wavelength shift. The Gaussian fitting resonant wavelength of dip2 undergoes intensity decreasing when the curvature increases from 0.94 to 2.10 m^{-1} [15].

4. Conclusion

In summary, this chapter introduces the optical fiber sensors based on ARROW. According to the working principle, the optical fiber sensors based on ARROW consist of the single layer, double layers, double resonators, and hybrid mechanism. Various optical fiber sensors based on ARROW have been introduced in this chapter with the aforementioned working principle, including the fiber optic vibration sensor, humidity sensor, strain sensor, temperature sensor, magnetic field sensor, biosensor, etc. The optical fiber sensors based on ARROW could enhance the interaction between the guided light and sensitive material, simplify the complexity of the sensor configuration, and increase the multifunctional performance of the fiber sensor. Especially, many long-standing challenges in the fiber optic sensor can be solved through the working principle of the ARROW, including the temperature cross-talk compensation, distribution localization, and optofluidic biosensing. In general, the optical fiber sensors based on ARROW have advantages, such as compact structure, high sensitivity, large dynamic range, and high stability, which appear to have potential applications in researches of structure health monitoring, oil exploiting, and biology detection.

Acknowledgements

The authors acknowledge the China National Key R&D Program (No. 2019YFA0706304) and the National Natural Science Foundation of China (Nos. 61835002, 61675033, 61727817, and 61601436).

Conflict of interest

The authors declare no conflicts of interest.

Appendix

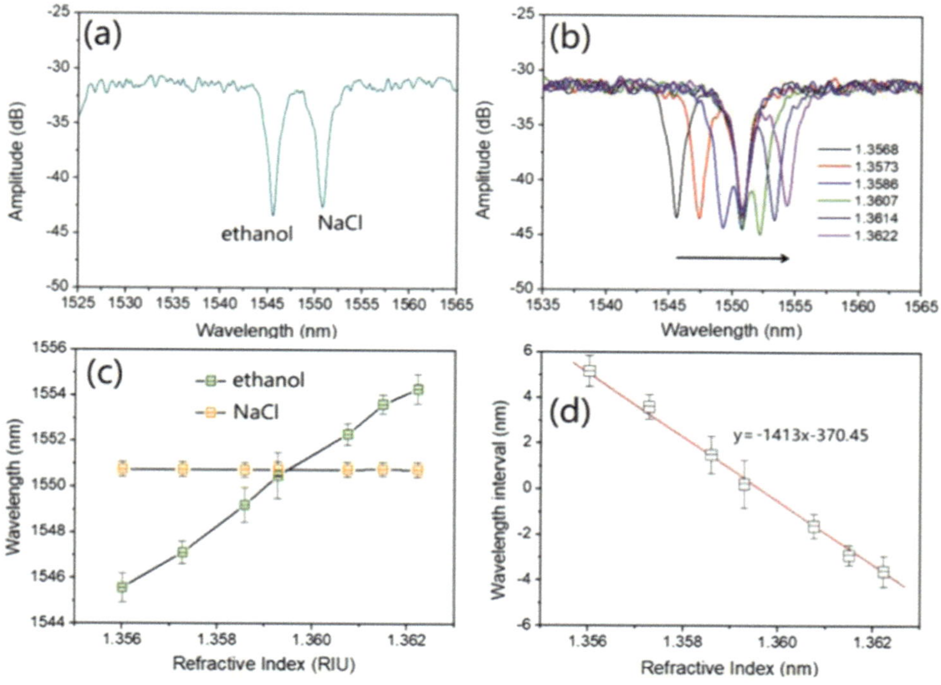

Figure 1A.
(a) Transmission spectrum of the dual-optofluidic waveguide ARROW biosensor, (b) wavelength shifts with different RI, (c) relationship between the wavelengths of resonance dips and RI, and (d) wavelength interval with different RIs [12].

Author details

Ran Gao[1*†] and Jiansen Ye[2†]

1 Advanced Research Institute of Multidisciplinary Science, Beijing Institute of Technology, Beijing, China

2 School of Information Engineering, Zhengzhou Institute of Technology, Zhengzhou, China

*Address all correspondence to: 6120190142@bit.edu.cn, forest929@163.com

† These authors are contributed equally to this work.

References

[1] Gao S-F, Wang Y-Y, Ding W, Jiang D-l, Gu S, Zhang X, et al. Hollow-core conjoined-tube negative-curvature fiber with ultralow loss. Nature Communications. 2018;**9**:1-6. DOI: 10.1038/s41467-018-05225-1

[2] Balciunas T et al. A strong-field driver in the single-cycle regime based on self-compression in a Kagome fiber. Nature Communications. 2015;**6**:6117. DOI: 10.1038/ncomms7117

[3] Zhang WL, Yu SF. Optical flip-flop using bistable vertical-cavity semiconductor optical amplifiers with anti-resonant reflecting optical waveguide. Journal of Lightwave Technology. 2009;**27**(21):4703-4710. DOI: 10.1109/JLT.2009.2025664

[4] Zhao P, Zhao Y, Bao H, Lu H. Mode-phase-difference photothermal spectroscopy for gas detection with an anti-resonant hollow-core optical fiber. Nature Communications. 2020;**11**(847): 1-8. DOI: 10.1038/s41467-020-14707-0

[5] Santos J S, Jr I M R, Cordeiro C M B, et al. Characterisation of a Nafion film by optical fiber Fabry-Pérot interferometry for humidity sensing. Sensors and Actuators B: Chemical. 2014;**196**(6):99-105. DOI:10.1016/j.snb.2014.01.101

[6] Litchinitser NM, Abeeluck AK, Headley C, et al. Antiresonant reflecting photonic crystal optical waveguides. Optics Letters. 2002;**27**(18):1592-1594. DOI: 10.1364/OL.27.001592

[7] Gao R, Lu D-F, Zhang M, Qi Z-M. Vibration sensor based on the resonance power leakage in a tapered capillary fiber. IEEE Sensors Journal. 2017;**17**(24): 8832-8837. DOI: 10.1109/JSEN.2017.2765891

[8] Gao R, Lua D-F, Chenga J, Jiangb Y, Jiangc L, Qi Z-M. Humidity sensor based on power leakage at resonance wavelengths ofa hollow core fiber coated with reduced graphene oxide. Sensors and Actuators B. 2016;**222**:618-624. DOI: 10.1016/j.snb.2015.08.108

[9] Liu S, Wang Y, Hou M, Guo J, Li Z, Lu P. Anti-resonant reflecting guidance in alcohol-filled hollow core photonic crystal fiber for sensing applications. Optics Express. 2013;**21**:31690-31697. DOI: 10.1364/OE.21.031690

[10] Gao R, Lu D, Cheng J, Qi Z-m. In-fiber double-layered resonator for high-sensitive strain sensing. IEEE Photonics Technology Letters. 2017;**29**:857-860. DOI: 10.1109/lpt.2017.2693218

[11] Li L, Xia L, Xie Z, Liu D. All-fiber Mach-Zehnder interferometers for sensing applications. Optics Express. 2012;**20**:11109-11120. DOI: 10.1364/OE.20.011109

[12] Gao R, Lu DF, Guo D, Xin XJ. Dual-optofluidic waveguide in-line fiber biosensor for the real-time label-free detection of interferon-gamma with temperature compensation. Optics Express. 2020;**28**:10491-1-5-4. DOI: 10.1264/OE.389766

[13] Gao R, Lu DF, Cheng J, Jiang Y, Qi ZM. Temperature-compensated fibre optic magnetic field sensor based on a self-referenced anti-resonant reflecting optical waveguide. Applied Physics Letters. 2017;**110**:131903-1311-5. DOI: 10.1063/1.4979412

[14] Gao R, Zhang Q, Xin X, Tian Q, Tian F. Chirped anti-resonant reflecting optical waveguide for the distributed sensing of pressure. Optics Letters. 2020;**45**:690-693. DOI: 10.1364/OL.382716

[15] Wenjun N, Ping L, Jiangshan Z, Chunyong Y, Xin F, Yuan S, et al. Single hole twin eccentric core fiber sensor

based on anti-resonant effect combined with inline Mach-Zehnder interferometer. Optics Express. 2017;**25**(11):12372-12380. DOI: 10.1364/OE.25.012372

[16] Duguay MA, Kokubun Y, Kuch TL. Anti-resonant reflecting optical waveguide in SiO2-Si multilayer structures. Applied Physics Letters. 1986;**49**:13-15. DOI: 10.1063/1.97085

[17] Bernini R, Campopiano S, Zeni L, et al. ARROW optical wave guides based sensors. Sensors and Actuators B. 2004;**100**:143-146. DOI: 10.1016/j.snb.2003.12.035

[18] Prieto F, Sepulveda B, Calle A, Llobera A, et al. Integrated Mach-Zehnder interferometer based on ARROW structures for biosensor applications. Sensors and Actuators B. 2003;**92**:151-158. DOI: 10.1016/S0925-4005(03)00257-0

[19] Litchinitser NM, Dunn SC, Usner B, et al. Resonances in microstructured optical waveguides. Optics Express. 2003;**11**(10):1243-1251. DOI: 10.1364/OE.11.001243

[20] Gao R, Jiang Y, Zhao Y. Magnetic field sensor based on anti-resonant reflecting guidance in the magnetic gel-coated hollow core fiber. Optics Letters. 2014;**39**:6293-6296. DOI: 10.1364/OL.39.006293

[21] Hou M, Zhu F, Wang Y, et al. Antiresonant reflecting guidance mechanism in hollow-core fiber for gas pressure sensing. Optics Express. 2016;**24**(24):27890-27896. DOI: 10.1364/OE.24.027890

[22] You B et al. Subwavelength film sensing based on terahertz antiresonant reflecting hollow waveguides. Optics Express. 2010;**18**(18):19353-19360. DOI: 10.1364/OE.18.019353

[23] Gao R, Jiang Y, Jiang L. Multi-phase-shifted helical long period fiber grating based temperature-insensitive optical twist sensor. Optics Express. 2013;**22**(13):15697-15709. DOI: 10.1364/OE.22.015697

[24] Zheltikov M. Ray-optic analysis of the (bio)sensing ability of ring-cladding hollow waveguides. Applied Optics. 2008;**47**(3):474-479. DOI: 10.1364/AO.47.000474

[25] G′erome F, Jamier R, Auguste JL, Humbert G, Blondy JM. Simplified hollow-core photonic crystal fiber. Optics Letters. 2010;**35**:1157-1159. DOI: 10.1364/OL.35.001157

[26] Couny F, Benabid F, Roberts PJ, Light PS, Raymer MG. Generation and photonic guidance of multioctave optical-frequency combs. Science. 2007;**318**:1118-1121. DOI: 10.1126/science.1149091

[27] You B, Lu J-Y, Liou J-H, Yu C-P, Chen H-Z, Liu T-A, et al. Subwavelength film sensing based on terahertz anti-resonant reflecting hollow waveguides. Optics Express. 2010;**18**:19353-19360. DOI: 10.1364/OE.18.019353

[28] Lai CH, You B, Lu JY, Liu TA, Peng JL, Sun CK, et al. Modal characteristics of antiresonant reflecting pipe waveguides for terahertz waveguiding. Optics Express. 2010;**18**:309-322. DOI: 10.1364/OE.22.008460

[29] Salceda-Delgado G, Van Newkirk A, Antonio-Lopez JE, Martinez-Rios A, Schülzgen A, Amezcua Correa R. Compact fiber-optic curvature sensor based on super-mode interference in a seven-core fiber. Optics Letters. 2015;**40**:1468-1471. DOI: 10.1364/OL.40.001468

[30] Sun B, Huang Y, Liu S, Wang C, He J, Liao C, et al. Asymmetrical in-fiber Mach-Zehnder interferometer for curvature measurement. Optics Express. 2015;**23**:14596-14602. DOI: 10.1364/OE.23.014596

[31] Zhao Y, Cai L, Li X. Temperature-insensitive optical fiber curvature sensor based on SMF-MMFTCSMF-MMF-SMF structure. IEEE Transactions on Instrumentation and Measurement. 2017;**66**:141-147. DOI: 10.1109/TIM.2016.2615479

[32] Amnon PYY. Photonics: optical electronics in modern communications. Vol. 849. New York: Oxford University Press; 2007. DOI: 10.1002/9781118755068.ch03

Chapter 3

Optical Waveguides Based on Sol-Gel Coatings

Helena Cristina Vasconcelos

Abstract

This chapter focuses on developing coatings for use as waveguides for integrated optics and photonics. Thin (or thick) films of silica-based inorganic materials and organic-inorganic hybrids can be easily obtained using the sol-gel and spin-coating method, followed by rapid thermal annealing to obtain dense films of good optical quality. The waveguide thermal, structural, and optical properties can be characterized using differential thermal analysis, X-ray diffraction, Fourier-transform infrared spectroscopy, Raman spectroscopy, scanning electron microscopy, and atomic force microscopy. Waveguides can be both doped with semiconductor microcrystallites and rare-earths for the development of optical devices, where light is confined to one or two dimensions (planar, channel or strip loaded).

Keywords: sol-gel, optical waveguides, coatings, spin-coating, silica, rare-earths, semiconductor microcrystallites

1. Introduction

The optical technology enables larger bandwidth over larger distances at enhanced power efficiency as compared to the use of electrical connections. In practice, this is based on the hybrid integration in one single substrate of many small individual components connected to each other by optical waveguides, an approach known as integrated optics (IO) (e.g., bring together on a single platform, filters, lenses, modulators, beam splitters, amplifiers, lasers, photodetectors, etc.). The basic idea behind IO is the management of light by waveguides and not by free space optical components like lenses and mirrors. Moreover, the success achieved in the long-distance transmission fiber in the 1960s has increased the attention toward the optical waveguides. Anderson, in 1965, used planar waveguides and optical circuit components for applications in the near infrared (IR) [1]. Later, in 1969, was first reported an experimental application of prims-film coupler and Miller first introduced the term IO [2]. Light can therefore be guided and confined by dielectric waveguides, such as those made of glass (SiO_2) deposited on quite a large number of substrates, but mainly silicon. The large waveguide cross-section allows easy coupling with optical fibers. Other hi-tech characteristics are their larger bandwidth, low energy requirements, and immunity to electromagnetic interference. Optical waveguides are thus the basic elements for confinement and transmission of light over various distances, ranging from a few μm, in IO, to several kilometers, in the transmission of optical fibers over long distances.

IntechOpen

The increasing use of optical communications, signal processing, and the development of new materials has created important advances in IO. The efficient use of the bandwidth offered by optical fibers also contributed to the success of this technology. The optical waveguides can be made of different materials. Each material has its own advantages and disadvantages. Therefore, the choice of a material to develop a new optical waveguide is usually the result of certain commitments. For example, considerable effort has been made to allow optical integration entirely using semiconductor materials (for example, GaAs/GaAlAs and InP/InGaAsP). However, its manufacture is not an easy task and requires complex epitaxial growth, usually by liquid phase epitaxy or metal-organic vapor epitaxy growth techniques. Unfortunately, these techniques do not offer the necessary manufacturing speed or the thicknesses required for the normal set-up of optical waveguides. Therefore, the use of glass (amorphous materials) has become widespread to develop planar waveguides (or slabs), as well as rapid deposition techniques (e.g., immersion or rotation of solutions to get glass by the sol-gel process). The sol-gel process has become a feasible method for fabricating thick films of desired composition to be carried out at speeds closer to those required in large-scale production. In addition, the use of silicon as a substrate also has numerous advantages due to the layout of silica-on-silicon (SiO_2/Si), one of the most important technologies in OI.

The arrival of low-loss silica fibers, whose refractive index perfectly matches the glass waveguides, contributed to making this material one of the most widely used in IO [3]. Currently, the sol-gel process offers great versatility because of the ease to functionalizing the glass structure with a wide variety of dopants. In fact, to provide all optical functions, from passive waveguides, electrical-driven modulators and switches, optical pumped active waveguides, optical-driven modulators and switches, and optical nonlinear devices, both optically active and optically nonlinear materials are needed. These materials can be respectively provided by two types of dopants: lasing species, such as rare-earths (RE) ions and semiconductor microcrystals. Hence, doping glass waveguides with Er^{3+} or Nd^{3+} ions allows glass to act as optical amplifier, while nonlinear properties can be introduced to silicate glasses by doping with semiconductor microcrystals and thus allowing to explore the Kerr effect [4]. Therefore, glasses can be used as optical waveguides, combining dielectric properties and providing specific functional responses for two main applications: as optical communication devices [5] and sensors [6].

2. Optical planar waveguides

A slab waveguide is similar to an optical fiber, except that it is in planar shape rather than cylindrical waveguide, where a low refractive index substrate contains on its surface a slab (or channel) of higher index material, along which light is guided by total internal reflection. The guiding material should also allow the linear and nonlinear optical properties to be varied in a straightforward way, over a wide range.

2.1 Linear dielectric waveguides

The most common structure of a dielectric waveguide is shown in **Figure 1**, which is made of three overlapping layers of dielectric materials [7]. The first layer (which is usually referred as the substrate) and the third layer (also named cladding) are both semi-infinite and have refractive indices n_1 and n_3, respectively, while the layer on the middle (guiding layer or core) has thickness d, and refractive index n_2. In order to ensure total internal reflection at each interface $n_2 > n_1$ and $n_2 > n_3$. It is illuminated at one end by a monochromatic light source.

A waveguide as shown in **Figure 1** can be made by depositing a high-index guiding layer onto a flat substrate were the cladding can be the air. Because of this asymmetric geometry, since $n_2 > n_1 > n_3$, this kind of waveguide is often described as an asymmetric slab waveguide (**Figure 1a**). However, if $n_1 = n_3$, the waveguide is symmetric (**Figure 1b**). Light is confined by the total internal reflection at the substrate and layer cladding interfaces; and the phenomenon can be analyzed by the ray optics approach. The ray of light experiencing successive total internal

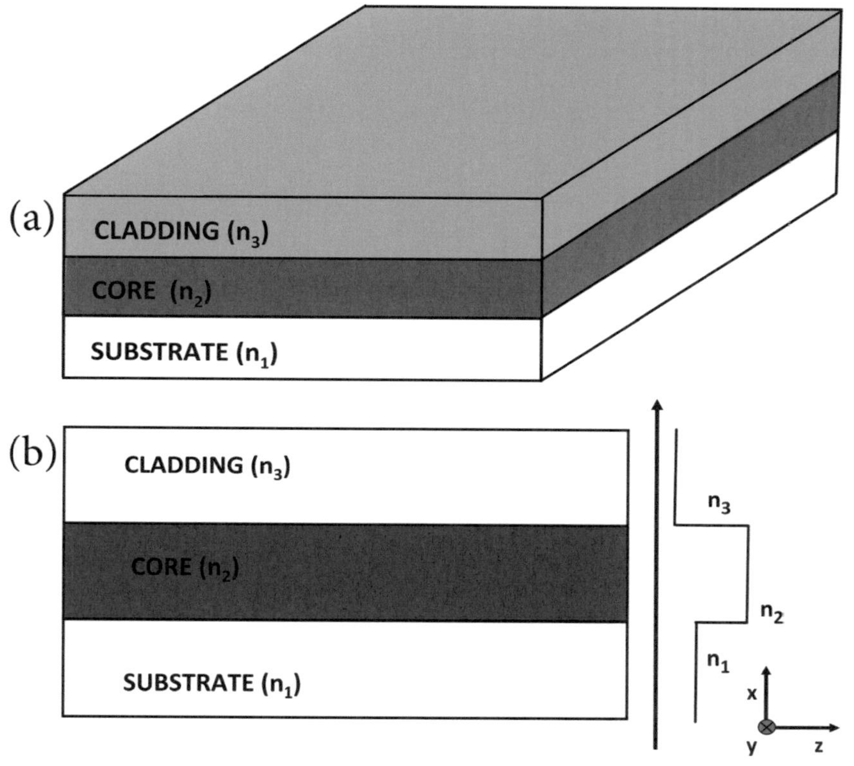

Figure 1.
Geometry of (a) an asymmetric slab waveguide ($n_2 > n_1 > n_3$), and (b) asymmetric slab waveguide with its refractive index profile.

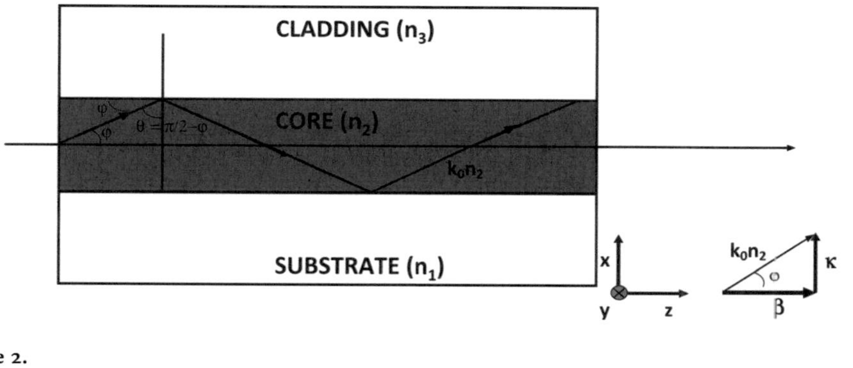

Figure 2.
Asymmetric slab waveguide with a propagating light ray in its core and geometric (vectorial) relationship between the propagation constants of dielectric waveguide.

reflections at the main interfaces will propagate continually along the central layer without loss [7]. This happens if the incident angle is small enough.

The propagation of signal within the guiding layer of planar waveguide is shown in **Figure 2**, where k_o is the plane wave propagation constant in free space (or free-space wavenumber). The free-space wavenumber can be expressed as a function of frequency and velocity, or the wavelength:

$$k_0 = \frac{\omega}{c} = \frac{2\pi f}{c} = \frac{2\pi}{\lambda} \quad (1)$$

From geometric optics, it is known that a ray of light will follow Snell's law while crossing the interfaces of the core with the two surrounding layers. In view of **Figure 2**, the condition for total internal reflection is given by [8]:

$$\sin^{-1}\left(\frac{n_1}{n_2}\right) < \theta < \sin^{-1}\left(\frac{n_3}{n_2}\right) \quad (2)$$

The light ray is traveling in the z-direction of an optically isotropic medium with constant internal reflection at the core-cladding interface (2–3) and the core-substrate interface (2–1). The ray must interfere constructively with itself to propagate successfully. Otherwise destructive interference will destroy the wave. The theory and equations of dielectric waveguide can be found in the literature [9]. However, here it is briefly highlighted just a few important mode equations. Assuming a TE polarization (electric field polarized in the z-direction), the propagation constants along z (longitudinal direction) and x (transverse direction) are, respectively, β and κ, giving by:

$$\beta = k_0 n_2 \cos\varphi \quad (3)$$

$$\kappa = k_0 n_2 \sin\varphi \quad (4)$$

At each interface, the light ray that undergoes total reflection suffers phase shifts, ϕ_{2-3} and ϕ_{2-1}, respectively, at the cladding and substrate, which depend on the refractive indices of the three layers (1, 2, and 3), and are equal to:

$$\phi_{2-3} = -2\tan^{-1}\left(\frac{\left(n_2^2\cos^2\varphi - n_3^2\right)^{\frac{1}{2}}}{n_2\sin\varphi}\right) = -2\tan^{-1}\left(\frac{\delta}{\kappa}\right) \quad (5)$$

$$\phi_{2-1} = -2\tan^{-1}\left(\frac{\left(n_2^2\cos^2\varphi - n_1^2\right)^{\frac{1}{2}}}{n_2\sin\varphi}\right) = -2\tan^{-1}\left(\frac{\gamma}{\kappa}\right) \quad (6)$$

where

$$\delta = \left(\beta^2 - k_0^2 n_3^2\right)^{1/2} \quad (7)$$

$$\gamma = \left(\beta^2 - k_0^2 n_1^2\right)^{1/2} \quad (8)$$

$$\kappa = \left(k_0^2 n_2^2 - \beta^2\right)^{1/2} \quad (9)$$

The phases $-2\phi_{2\text{-}3}$ and $-2\phi_{2\text{-}1}$, represent the Goos-Hänchen shifts [10]. From the solution of Maxwell's equations, it can be demonstrated that only certain discrete values of β are allowed [11]. Therefore, there are only a limited number of guided modes and the condition for being a guided mode is that β must be within two limits (up and down):

$$k_0 n_1 < \beta < k_0 n_2 \tag{10}$$

To avoid destructive interference as the light travel through the guide, the total phase shift for a ray that travels from the 2-3 interface to the 2-1 interface and back again must be a multiple of 2π. This leads to the condition [12]:

$$2k_0 n_2 d \sin\varphi - 2\phi_{2-3} - 2\phi_{2-1} = 2m\pi \tag{11}$$

where d is the thickness of the core (layer 2) and m is the mode number, which is a positive integer (0, 1, 2, ...). The fundamental mode corresponds to $m = 0$ and also to the lowest thickness (d_{min}) of the waveguide. In other words, for the fundamental TE mode to be supported by a planar waveguide, the thickness of the waveguide must by equal or larger than d_{min}. The dispersion relation for the slab waveguide [7] (or eigenvalue equation of the TE fundamental mode supported by the waveguide) is given by:

$$\tan(d\kappa) = \frac{\kappa(\delta+\gamma)}{\kappa^2 - \gamma\delta} \tag{12}$$

From Eq. (11), it is possible to obtain the minimum thickness for an asymmetric planar slab waveguide, which is given by:

$$d_{min} = \kappa^{-1}\left(\tan^{-1}\frac{\kappa(\delta+\gamma)}{\kappa^2 - \gamma\delta}\right) \tag{13}$$

From Eq. (9), if $\beta \to k_0 n_2$ (up limit), then $k = 0$ (as shown by the vectorial relationship of **Figure 2**), which correspond at the cut-off of the mode. Thus, in this situation, the waveguide thickness tends to infinite. In practical situations, this means that the waveguide thickness would become very large. Otherwise, if $\beta \to k_0 n_1$ (lower limit), then $\gamma = 0$ as shown by Eq. (8). Therefore, Eqs. (7) and (9) can be rewritten, respectively as:

$$\delta = k_0\left(n_1^2 - n_3^2\right)^{1/2} \tag{14}$$

$$\kappa = k_0\left(n_2^2 - n_1^2\right)^{1/2} \tag{15}$$

Substituting both Eqs. (14) and (15) in Eq. (13), after some rearrangement, it is then obtained the waveguide minimum thickness for asymmetric case:

$$d_{min} = \frac{\tan^{-1}\left(\left(n_1^2 - n_3^2\right)/\left(n_2^2 - n_1^2\right)\right)^{1/2}}{k_0\left(n_2^2 - n_1^2\right)^{1/2}} \tag{16}$$

For a single-mode operation at 1.5 μm wavelength, a guiding film with a refractive index equal to 1.52, deposited onto a silica glass substrate ($n_1 = 1.46$) surrounding by air ($n_3 = 1$), should have a thickness higher than 670 nm. Since the guide is asymmetric, then the maximum wavelength that can be guided to a guide layer thickness $d > d_{min}$ is given by [13]:

$$\lambda_{max} = \frac{2\pi d\left(n_2^2 - n_1^2\right)^{1/2}}{\tan^{-1}\left(\left(n_1^2 - n_3^2\right)/\left(n_2^2 - n_1^2\right)\right)^{1/2}} \tag{17}$$

This applies to transverse electric (TE) polarization, but a similar condition applies to transverse magnetic (TM) polarization [13]. Because of the absence of cut-off, symmetric waveguides ($n_1 = n_3$) permit the propagation of the guiding mode in a layer of very small thickness. In addition, symmetric waveguides need a cover layer of similar refractive index to that of the substrate. The cover layer serves as the protective cladding for the gain layer. The presence of the cover layer also increases the confinement factor of the guiding modes [9], resulting in reduced modal loss and hence pump threshold reduction.

The typical way to achieve modal analysis in a planar optical waveguide is by means of *m*-line spectroscopy, by a prism coupler, where light is introduced into a waveguide by means of evanescent wave coupling through a high index prism of a material.

However, there are three major mechanisms of losses in waveguides, and thereby the energy of guided modes is often attenuated by losses due to mechanisms of absorption (owing to photons destruction in materials), radiation (owing to waveguide bending or curvatures), and scattering (owing to surface roughness) [11]. Absorption loss is mainly due to material contaminations while scattering loss in dielectric planar waveguides is due in part by internal defects such as pores or crystalline defects, but mainly to surface roughness. Rayleigh and Mie scattering are the principal mechanisms of scattering loss and are both linear occurrences since they do not change the wavelength of the scattered light (contrary to what happens in the non-linear scattering caused by Raman and Brillouin mechanisms). Rayleigh scattering is inherent to glass and amorphous materials and is caused by the lack of homogeneity of the refractive index over small distances when compared to the wavelength of light. On the other hand, Mie's dispersion is due to the presence of defects and occurs at distances of the order of magnitude of the wavelength of the input light [11].

2.2 Nonlinear optics

Whenever a beam of light (of low intensity) propagates through a dielectric material, the nuclei and associated electrons of the atoms in the material form electric dipoles which oscillate at the same frequency as the incident beam, thus giving a linear relation between the polarization density and the electric field, $P = \varepsilon_0 \chi E$, where ε_0 is the permittivity of free space and χ is the electric susceptibility of the material [4]. For high intensity light, such as laser light, the amplitude of response of the atoms is no longer linear. Therefore, to the polarization density should be added higher order nonlinear terms and P represented as a power series expansion of the applied electric field E of the light:

$$P = \varepsilon_0 \left(\chi^{(1)} E^1 + \chi^{(2)} E^2 + \chi^{(3)} E^3 + \chi^{(4)} E^4 + \ldots \right. \tag{18}$$

were $\chi^{(1)}$, $\chi^{(2)}$, and $\chi^{(3)}$ are, respectively, the linear, second order and third-order susceptibilities. In *centrosymmetric* materials, the second order susceptibility is however absent, $\chi^{(2)} = 0$, and so the lowest order nonlinearity is the third-order:

$$P_{nl} = \varepsilon_0 \chi^{(3)} E^3 \tag{19}$$

P_{nl} become the dominant nonlinear component in Eq. (18). Due to $\chi^{(3)}$ nonlinearity, the pump wave generates a nonlinear polarization (output wave) which oscillates at frequency three times higher than the input frequency. The optical nonlinearity $\chi^{(3)}$ is a tensor with various coefficients being associated with various effects, including the Kerr susceptibility $\chi^{(3)}$ (3ω). The Kerr effect is the intensity dependence of refractive index, which can be expressed by a nonlinear index coefficient n_2 (the Kerr coefficient) proportional to $\chi^{(3)}$. Thus, the total index n is given by:

$$n = n_0 + n_2 I \tag{20}$$

where n_0 is the linear refractive index and I is the optical intensity in W/cm^2 (thus n_2 will have units of cm^2/W).

Nonlinear optical materials are needed for communications components [14]. The knowledge of the nonlinear refractive index n_2 is essential to the analysis and use of nonlinear devices. While silica glass has exceptional optical characteristics for waveguide proposes, they very low nonlinear coefficient turn it in a material unsuitable for nonlinear applications. However, strong nonlinearities responses can be obtained by the addition of, for example, semiconductor microcrystallites to the silica host. What give to these microcrystals such capacity is that the nonlinear effects can be significantly enhanced compared with those of the bulk semiconductors. Hence, when the size of the microcrystallites decreases to nanometric sizes, its electronic properties start to change from bulk state properties to that of a quantum dot state [15]. The essential change is that of the increase of excitonic level relative to that of bulk semiconductor (blue shift) as the radius of microcrystallites semiconductor is reduced. Band-filling models have been successfully used to explain the maximum enhancement for crystal sizes in the vicinity of the excitonic Bohr radius (~10–20 Å) [15]. For smaller crystals, size quantization effects dominate and the electron-hole Coulomb attraction becomes insignificant.

3. Sol-gel process for the fabrication of waveguides

Glasses are suitable materials for application of passive, active, and nonlinear optical devices, due to their characteristic physical and chemical properties, in particular their optical features, such as exceptional transparency and high threshold to optical damage. In particular, silica glasses are very attractive because they match perfectly with the optical fibers and thus present small coupling losses; high thermal, chemical, and mechanical stability. Moreover, silica glass exhibits very low expansion coefficient over silicon which allows the deposition of thick buffer films without failure, which is a crucial feature layer on silicon for waveguide proposes. Of particular importance in the field of photonic glasses is the SiO_2/Si technology, where silica glass-based waveguides can be fabricated using several techniques. Flame hydrolysis deposition (FHD) and chemical vapor deposition (CVD) provide excellent quality thin films and have been extensively investigated [16, 17]. An alternative approach to these production methods is the sol-gel process, which offers many advantages in terms of avoiding complex equipment and high cost procedures [18].

3.1 Sol-gel process to produce glass and ceramic materials

The sol-gel process is a method for producing glass and ceramic materials by low temperature polymerization reactions based on hydrolysis and condensation of alkoxides, whose general formula is $M(OR)_n$, where M is a metal atom and R is one alkyl group that allow to make various glassy materials with remarkable optical or photonic properties. This process allows the preparation of high-purity and homogeneous materials and results in the preparation of different kinds of shapes, like bulk solids, fibers, film waveguides, and coatings for device applications with a great diversity in material composition and structures which is important for control of the functionality of IO devices [19]. The main advantage of sol-gel process is that of allowing thin film deposition, providing a great versatility in the development of devices with special optical functions, particularly materials with a strict chemical control of the processing parameters. This flexibility allows diverse modifications to the host glass and the addition of a wide variety of dopants, thus greatly increasing the range of functions and applications of the resulting components. In the fabrication of sol-gel waveguides, a careful control from the beginning of the process is required in order to enhance the properties of these devices, starting from a suitable choice of the chemical precursors up to a correct final annealing. Through the control of the viscosity of the deposition solution, different coating processes like dipping, spraying, and spinning can be used. The strict chemical control and high purity of the starting reagents, mixed in the liquid state, lead to homogenous gels with low impurity levels, which is especially adequate for photonic technologies.

Dopants in the form of salts or alkoxides can be added to the precursor solution (sol) in order to prepare doped photonic inorganic materials. However, a limitation for application of sol-gel materials into optical devices is the drawback imposed by the maximum achievable crack-free sol-gel glass thickness [20]. SiO_2/Si technology compatible with single mode fiber for 1.55 μm, the most widely used wavelength in optical communication, requires waveguides typically greater than 1 μm in thickness and thus the need to employ a multistep deposition procedure (each layer being 150 to ~300 nm thick). To overcome this obstacle, rapid thermal annealing has been used to fabricate films through the deposition of iterative cycles of deposition, achieving the total required thickness [21].

Almost all of the sol-gel planar waveguides are fabricated on asymmetric structure with a substrate of silica glass (or a silicon with a buffer layer) and air as cladding layer.

3.2 Sol-gel process of film formation

The basis of SiO_2/Si technology is to deposit silica layers on silicon substrates, to define single-mode channel waveguides in this material, and to couple these guides to external fibers. In the absence of a taper or similar component, the mode fields of the channel guides should match those of single-mode silica fiber as closely as possible, in order to limit coupling losses. However, use of more strongly confining guides allows reduced bend radii and therefore lowers component size and also concentrates pump power, to great advantage in amplifiers and nonlinear devices [22]. The SiO_2-based glass layers can be easily deposited by spin coating using the sol-gel process. A two-step process based on the sol-gel process can be successful used for fabricating amorphous dielectric layers [21]:

1. formation of a stable suspension of particles within a liquid (the sol) and further processing of the sol to form mainly Si-O-Si bonds thought condensation reactions (aging sol);

2. spin-coating of the aging sol and rapid thermal annealing (multiple cycles) to form a thick and dense glassy film.

The first step involves the use of liquid metal alkoxide precursors of the desired glass composition. A common precursor for SiO_2 is tetraethoxysilane (TEOS), a liquid at room temperature insoluble in water but soluble in ethanol. This precursor is hydrolyzed by addition of water to the solution, which allows condensation to begin to form the Si-O-Si bonds of the glass. However, other precursors may be used (added) following the same steps. For example, the process for preparation of SiO_2-TiO_2 solutions uses, for example, titanium isopropoxide IV (TPOT) as TiO_2 precursor [23]. After an aging time of ~16 h (**Figure 3**), the suspension is spun at typically 2500 rpm, during 30 s, upon which the solvent evaporates and the stages of solidification and drying are greatly accelerated for a few seconds, giving a thin layer of ~300 nm.

The gel must be heated to remove residual organics and to complete densification but this stage induces large tensile stresses [20–24]. The gel is therefore a three-dimensionally-linked solid network with liquid (solvents) filling the pores, which are interconnected in the wet gel state. The removal of solvent during drying proceeds simultaneously with condensation reactions and solidification. As the solvent is evaporated, a competing process leads to capillary pressure and the stresses induced by shrinkage increases. Thus, cracking may happen. It has been shown that the thickness of a single layer SiO_2 is limited to approximately 0.7 μm with the sol-gel method [25]. At thickness greater than 0.7 μm, the intrinsic stress and mismatch of thermal expansion coefficients between the sol-gel SiO_2 layer and the substrate causes cracks and peeling. Moreover, the constrained one-dimensional shrinkage due to the substrate creates tensile stress within the sol-gel layer [20–24], which must be annealed out to prevent cracking. Iteration of this process has been successful used to produce a multi-micron film [21]. Cracks and pores are almost inevitable in sol-gel derived films because of organic components, which must be eliminated during heat treatment. Thus, residual pores still remain in sol-gel waveguides after the removal of organic residues at high temperatures (**Figure 4**). By varying the precursor ratios in the sol, glasses of different composition, and thus refractive index, can be achieved; for example, SiO_2/TiO_2 of differing molar ratios allows a wide index range from 1.46 to over 1.6. In particular, for waveguide devices, a bilayer is formed: a lower index buffer (usually a SiO_2 layer), which must be sufficiently thick to prevent leakage of guided light into the substrate, and a higher index guiding layer (e.g., SiO_2-MO_2 (M = Ti, Zr, Hf) or SiO_2-HfO_2-MO_2 (M = Ti, Zr) materials [26]). The thickness and refractive index of films can be measured using a profilometer and an ellipsometer, respectively, as a function of the sol composition and preparation conditions (water, catalysts, and solvents amounts). The refractive index, whose control is necessary for the films to operate as planar waveguides in the SiO_2/Si technology, is usually tailored by doping the SiO_2 material, for example, with TiO_2 (or other, e.g., GeO_2). However, due to the

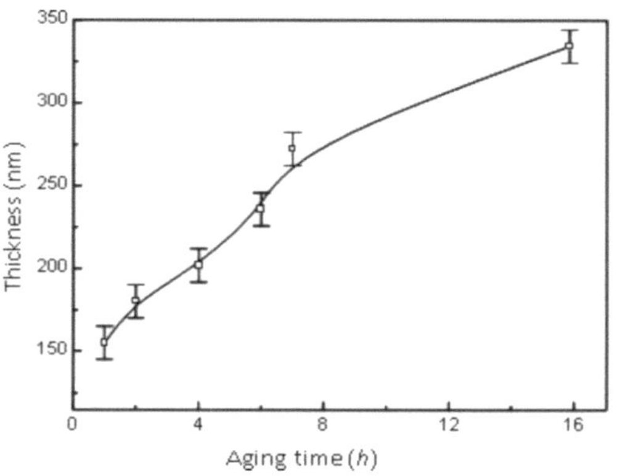

Figure 3.
Effect of aging time on the thickness of spin-coated sol-gel $90SiO_2$-$10TiO_2$ (mol%) films.

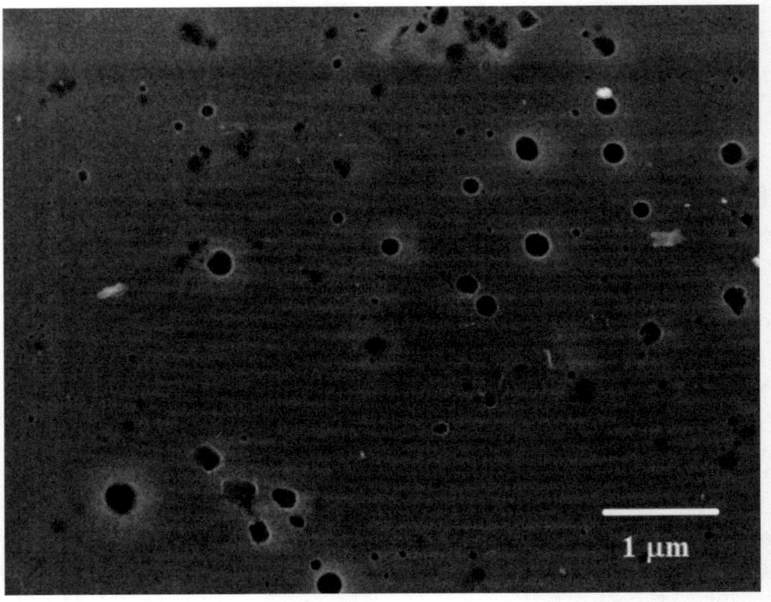

Figure 4.
A typical sol-gel glassy thin film. The dark circular regions represent the pores.

different rates of hydrolysis of the different precursor alkoxides, the compositional homogeneity of the derived material must be investigated by transmission electron microscopy (TEM) and Auger (or X-ray photoelectron spectroscopy (XPS)) analysis. It is also important to follow the evolution of the densification process through refractive index measurements but also directly measuring the surface area and porosity of the films after specific heat treatments, using the Brunauer-Emmett-Teller (BET) Surface Area Analysis and Barrett-Joyner-Halenda (BJH) Pore Size and Volume Analysis.

The guiding layers may also be doped with a functional material; for example, rare-earths (RE) can be added to the sol through a variety of precursors.

Also nonlinear properties can be introduced in silicate glasses by doping with semiconductor nanocrystals.

A modification of the most common planar bilayer guiding structure can be achieved by etching the layer of high index (n_2) into ridges (**Figure 5a**), which can then be melted (reflowed) to improve shape and surface quality and then buried under a top cladding glass (**Figure 5b**) of lower index (n_1). In an alternative approach, light can be guided in the high index layer by shallow ridges on the upper cladding layer; this is called a strip loaded structure (**Figure 5c**). In this case, etching and reflow at the guiding layer is avoided. The production of silica channel waveguides frequently employs photolithography; the mask for etching the strips can be the photoresist itself or a metallic film and reactive ion etching (RIE), which provides high anisotropy. Passive ridge waveguides were deposited on silicon by a solvent-assisted lithographic process incorporating simple mask technology and photosensitive sol-gel-derived glasses [27]. This approach uses direct writing of photosensitive glass with UV light.

With the synthesis of organic-inorganic films (ormosils), it is possible to obtain hybrid materials with film thickness much higher than 1 µm, which fit the requirements of optical waveguiding. For the preparation of these materials, TEOS (or other alkoxide) can be mixed together with organosilane polymers (e.g., polydimethylsiloxane (PDMS) or 3-(trimethoxysilyl)-propylmethacrylate

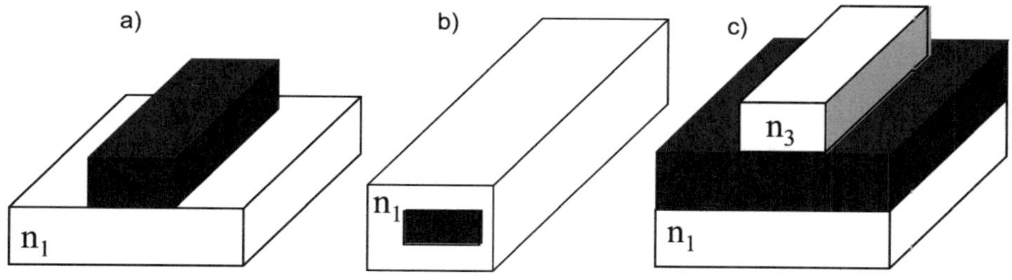

Figure 5.
Typical configuration of channel waveguides: (a) ridge waveguide, (b) buried waveguide, and (c) strip-loaded waveguide.

(TMSPM)), which keep trapped inside the oxide gel network during the hydrolysis and condensation reactions of TEOS [28]. In that case, the organic groups can further be used as oligomers since their backbone structures are similar to that of silica matrix. In addition, these groups can also improve physical, chemical, and mechanical properties of the hybrid materials, and the modification of inorganic network structures with organic groups allow the isomerization of organic photoactive molecules as compared to inorganic glasses [29]. It has been reported the use of two different silanes: γ-glycidyloxypropyltrimethoxysilane (GLYMO) and 3-(trimethoxysilyl)-propylmethacrylate (TMSPM) [30] for the preparation of TiO_2/ormosil waveguide films by the sol-gel method.

In view of the above discussion, it can be established that there are three main characteristics required for sol-gel waveguides:

1. mode shapes suitable for coupling to single-mode fibers (by absolute refractive index, refractive index difference and geometry);

2. compatibility with dopants (e.g., salts, alkoxides, organosilane polymers, etc.);

3. Compatibility with further processing steps, namely photonic materials with specific functional responses (e.g., optical amplification and nonlinear properties).

Some of the challenges still pending currently are: reduction of propagation losses for a planar waveguide to less than 1 dB/cm (although the short path lengths involved in integrated circuits make optical losses less of a problem), reduction coupling losses between optical fibers and integrated optical waveguides, and difficulty in directing light around sharp curves due to system miniaturization. Key future goals are to develop constituent functions required for fully integrated amplifiers; in particular, a tapered coupler for efficient power transfer between fibers and strong confinement guides, and a wavelength selective filter for coupling pump and signal wavelengths. A key issue is the need to achieve minimal loss with high confinement waveguides.

4. Doping silicate waveguides

4.1 Rare-earths doping for amplification

Optical fibers doped with rare-earth (RE) ions have contributed to the fast development of optical telecommunications [31]. Adding amplification to passive functions of simple waveguides could allow low cost and compact amplifiers, through on chip integration of multiple components. Hence, developing RE-doped layers for optically pumped amplifiers is a key challenge. This can be provided in glasses doped with Nd^{3+} and Er^{3+}. Both have been widely used in fiber amplifiers and lasers, but in these cases, the gain was less important due to the very low losses in the fibers and the long path lengths that can easily be achieved.

In order to reduce the fiber and waveguide path lengths, the use of relatively large concentrations of RE ions of high quantum efficiency is required and, therefore owning long metastable level lifetimes. Nd^{3+} can be added to the sol as a nitrate and then converted to the oxide form by heat treatment. However, Nd_2O_3 is immiscible in silica glass for amounts higher than 0.5 wt% (by conventional melt-casting), and therefore it forms aggregates or clusters [32]. These clusters cause non-radiative deexcitation processes which greatly lower the fluorescence lifetime and thus the gain of the material. Hence, these lifetimes are seriously reduced by concentration quenching (or clustering) due to neighbor ion interactions in insufficient dilute systems. In addition, non-radiative quenching phenomena may also be caused by residual OH groups or by multiphonon relaxation, which are strongly favored in high vibrational energy matrices like silica glasses [33, 34]. Similar behavior is found for high levels of Er^{3+} ions in silicate glasses, also due to their poor solubility on that material. The ion dispersion is dependent on dopant concentration and on the solubility of dopant in the host material, which fortunately in the case of sol-gel glasses, can be improved. Hence, sol-gel-derived glasses are especially suitable to obtain homogenous compositions with a good dispersion of RE ions. Co-dopants (P_2O_5 or Al_2O_3) are often used to increase the solubility of the Er^{3+} doping. Er^{3+} is particularly interesting as it exhibits a laser transition at 1.55 μm and, thus, can provide amplification in the third optical communications window [31]. The energy-level structure of Er^{3+} allows pumping of the $^4I_{11/2}$ excited-state multiplet around 980 nm by semiconductor lasers and the $^4I_{13/2}$ excited-state multiplet is populated by subsequent multiphonon relaxation. Radiative relaxations of the $^4I_{13/2}$ to the $^4I_{15/2}$ ground-state multiplet may then provide gain near 1.55 μm [31].

However, the optical properties of RE-doped materials strongly depend on the host system, the dopant concentration, and the thermal background.

The method of incorporation of Er^{3+} consists in the dissolution of erbium salts (erbium (III) nitrate or chloride pentahydrate) into the sol, followed by aging and then coating. In this case, the ions will favor non-bridging oxygen sites in the gel as the solvent is evaporated. Introducing Er^{3+} in ionic form as dissolved salts is less effective for its incorporation into the sol network, and consequently, as the solvent is evaporated during film formation, some precipitation may occur, resulting in a segregated film. The use of an erbium alkoxide precursor favors the condensation reactions between the Er^{3+} ions and the host material in the sol and thus form a homogeneous structure with a strong possibility of avoid cracks formation during the densification process. In some cases, small amounts of TiO_2 or P_2O_5 nucleating agents are also added to the sol with the purpose of causing controlled nucleation and crystallization of erbium-containing nanocrystalites. This approach was successfully tested in the case of Er^{3+}:SiO_2-TiO_2 and Er^{3+}:SiO_2-TiO_2-P_2O_5 sol-gel glasses, where $Er_2Ti_2O_7$ and $ErPO_4$ nanocrystalline phases, respectively, with good spectroscopic properties, were obtained [35]. The precipitation of such nanocrystallites allows the control distribution of the Er^{3+} ion in the matrix, reducing their clustering effect and maximizing their quantum efficiency. To characterize the active properties of the waveguides, measurements of the fluorescence spectra and the lifetime of the erbium $^4I_{13/2}$ metastable level can be performed.

The surface structure (e.g., surface roughness, nanocrystallites, porosity, etc.) can be investigated by atomic force microscopy (AFM) and scanning electron microscopy (SEM) (**Figure 4**). Clusters can be analyzed by grazing angle X-ray diffraction or very low frequency Raman scattering. To investigate the glass structure, it can be used X-ray absorption fine structure spectroscopy (EXAFS). This tool provides a good probe of RE nearest-neighbors and reveals, for example, where the Er^{3+} ions are well dispersed simple by Er-O bonds relative intensity [36]. A previous study reported from XPS analysis that Nd^{3+} may be incorporated in the P_2O_5 co-doped $80SiO_2$-$20TiO_2$ films at a higher concentration level than Er^{3+} in similar Al_2O_3 co-doped films [37]. Crystallization can, however, be investigated by X-ray diffraction (XRD) and crystal size can be estimated by the Scherrer equation. In particular, nanocrystallites are only seen at grazing incidence (0.4°), suggesting their main concentration at the surface [38]. In order to obtain concentration gradients through the total film thickness, Rutherford Backscattering Spectroscopy (RBS) can be used.

Fluorescence lifetime can be measured on multilayer planar waveguides, by launching diode laser pump light (796 nm for Nd^{3+}, 976 nm for Er^{3+}) into the cleaved input end of the waveguide. The fluorescence is collected at the output end and then focused onto a fast photodetector for further analyzes with a digital oscilloscope.

Since the formation of erbium containing nanocrystals in glass results in an increase for the 1.55 μm fluorescence (up to 9 ms) [35], the determination of characteristic temperatures such as glass transition (Tg), onset crystallization temperature (Tx), crystallization temperature (Tc), onset melting temperature (Txf), and melting temperature (Tf) is important in estimating the thermal stability of a glass and its susceptibility to temperature-induced changes during the film annealing. A study based on DTA analysis revealed that anatase crystallites appear as the primary crystalline phase in SiO_2-TiO_2-P_2O_5 glasses, while cristobalite precipitates near 1000°C [39]. Thus, the characterization of these temperatures using differential thermal analysis (DTA) technique would be a great help to define the thermal treatment of the sol-gel glasses derived waveguides.

4.2 Semiconductor microcrystallites doping for nonlinear optics

Semiconductor quantum dots dopants (Cds, Pbs) can extend the functionality of passive glassy waveguides toward possible applications in nonlinear guided wave devices (nonlinear directional coupler, Mach-Zehnder interferometer). Significant enhancement of $\chi(3\omega)$ due to quantum confinement has been demonstrated in glass doped with CdS_xSe_{1-x} and CuCl microcrystals [40]. In both cases, the large band gap means that the non-linearity can only be exploited at visible or near-UV. Recently, narrow-gap dopants have been fabricated at near-IR wavelengths [41].

Two main routes have been followed in the development of such materials, namely the addition and reaction of the dopants inside the initial sol-gel solution, by chemically controlling the size of the particles (sol-doping), and the impregnation or exposition of a nanoporous sol-gel film to the dopants (pore doping), where the crystal size is limited by the pore size. The pore doping has been successfully applied for over 20 years [42] to prepare films for integrated optics applications. In the sol doping, $Cd(NO_3)_2$ and $Pb(NO_3)_2$ is mixed in aqueous solution since both nitrates have very high solubilities and the resulting solution absorbed into a dry porous SiO_2 layer. The excess liquid is then spun off and a flux of H_2Te gas is forced to pass over the layer in order to achieve, for example, Cd1-xPbxTe microcrystallites. Otherwise, if the intention is rather precipitate CdS or PbS, H_2S can be used as reactive gas. The samples are then dried in vacuum and properly annealed. To prove the existence of microcrystallites precipitates, TEM and grazing incidence XRD is often employed to verify their presence in the host glass. Moreover, semiconductor microcrystallites are known to exhibit an absorption edge blue shifted (toward shorter wavelength) with respect to the bulk semiconductor. This shift is a typical behavior of microcrystallites that can be analyzed by UV-visible absorption spectroscopy in order to calculate their quantum size [42]. Finally, the Z-scan technique is a popular method for testing optical nonlinearities of bulk materials, namely the non-linear index n_2 (Kerr coefficient) and the non-linear absorption coefficient $\Delta\alpha$. However, these techniques are inappropriate for nonlinear measurements in planar optical waveguides due to its low thickness. Therefore, nonlinear m-line spectroscopy is a suitable alternative based on the analysis of the nonlinear change in the shape of the dark line associated with the excitation of guided waves [43]. This is an easy technique for the determination of Kerr properties of thin films, which are the basis of optical waveguides [43]. The third-order optical nonlinearity (Kerr effect) of CsS-doped SiO_2/Si planar waveguide was measured using an m-line technique, yielding a value of -5×10^{-9} cm^2/kW [44].

5. Conclusions

Silica glass (by sol-gel synthesis) is one of the most widely used materials in the production of optical waveguides, due to its many advantages (versatility, low cost manufacturing, chemical resistance, and low synthesis temperature). The chemical kinetics of the conversion reaction of TEOS to SiO_2 (or other glass compositions) is very sensitive to the physical and chemical parameters of the process (e.g., water content, type of catalyst, and aging). However, its versatility is one of its main advantages because allows the production of a diversity of compositions and structures. The spin-coating process allows homogeneous and thick films to be deposited, which perfectly match the physical requirements of planar optical waveguides and that of SiO_2/Si technology. However, silicate glasses exhibit some disadvantages, such as high phonon energy and low luminescent ion solubility, which affect the quantum efficiency or luminescent ion emission bandwidth. Therefore, some strategies must be taken to improve their optical behavior. In fact, by slightly modifying the composition of the silica glass, we are able to control the refractive index and achieve not only passive waveguides but also the active ones, with tailored properties to many applications. Moreover, it is also possible to take advantage of the pores in order to optimize the precipitation of semiconductor microparticles and benefit of all the linear and nonlinear features of optical devices. Nowadays, the requirements for photonic components in communication systems have increased and become more demanding. While other technologies have made constant improvements, the potential advantages of the sol-gel approach remain valid and have now been plainly demonstrated. Although many of the procedures implemented are more than 25 years old, it is certain that, without all these fundamentals, the present development in the field of optical waveguides would not be the same.

Author details

Helena Cristina Vasconcelos[1,2]

1 Faculty of Sciences and Technology, Azores University, Ponta Delgada, Portugal

2 Faculty of Science and Technology, Department of Physics, Centre of Physics and Technological Research (CEFITEC), New University of Lisbon, Caparica, Portugal

*Address all correspondence to: helena.cs.vasconcelos@uac.pt

References

[1] Wilson J, Hawkes JFB. Optoelectronics: An Introduction. Taipei, Taiwan: Pearson Education Taiwan Ltd; 2010

[2] Miller SE. Integrated optics: An introduction. Bell System Technical Journal. 1969;**48**:2059-2069. DOI: 10.1002/j.1538-7305.1969.tb01165.x

[3] Osterberg H, Smith LW. Transmission of optical energy along surfaces: Part II, inhomogeneous media. Journal of the Optical Society of America. 1964;**54**:1078-1084. DOI: 10.1364/JOSA.54.001078

[4] Boyd RW. Nonlinear Optics. 3rd ed. USA: Academic Press, Inc.; 2008

[5] Yeatman EM. Sol-gel fabrication for optical communication components: Prospects and progress, In: Proc. SPIE. 1997;**68**:119. DOI: 10.1117/12.279834

[6] MacCraith BD, McDonagh C, McEvoy AK, Butler T, O'Keeffe G, Murphy V. Optical chemical sensors based on sol-gel materials: Recent advances and critical issues. Journal of Sol-Gel Science and Technology. 1997;**8** (1053). DOI: 10.1023/A:1018338426081

[7] Marcuse D. Chapter 1 - The Asymmetric Slab Waveguide. In: Marcuse D, editor. Theory of Dielectric Optical Waveguides. 2nd ed. Academic Press; 1991. pp. 1-59. DOI: 10.1016/B978-0-12-470951-5.50007-X

[8] Grivas C. Optically pumped planar waveguide lasers, part I: Fundamentals and fabrication techniques. Progress in Quantum Electronics. 2011;**35**(6): 159-239. DOI: 10.1016/j.pquantelec.2011.05.002

[9] Okamoto K. Fundamentals of Optical Waveguides. San Diego: Academic Press; 2000

[10] Wild WJ, Giles CL. Goos-Hänchen shifts from absorbing media. Physical Review A. 1982;**25**(4):2099-2101

[11] Hunsperger RG. Integrated Optics: Theory and Technology. 6th. ed. Springer Publishing Company, Incorporated; 2009

[12] Righini GC, Klein L, et al., editors. Handbook of sol-gel science and technology. In: Characterization of Sol–Gel Thin-Film Waveguides. DOI: 10.1007/978-3-319-19454-7_46-1

[13] Yeatman EM. Thin-film optical waveguides. In: Aegerter MA, Mennig M, editors. Sol-Gel Technologies for Glass Producers and Users. Boston, MA: Springer; 2004. DOI: 10.1007/978-0-387-88953-5_42

[14] Yoshiki W, Tanabe T. All-optical switching using Kerr effect in a silica toroid microcavity. Optics Express. 2014;**22**:24332-24341

[15] Semiconductor Quantum Dots, Bányai L, Koch SW. World Scientific Series on Atomic, Molecular and Optical Physics; 1993

[16] Silica-on-Silicon Integrated Optics. Available from: http://www3.imperial.ac.uk/pls/portallive/docs/1/2475919.PDF

[17] Giancarlo C, Righini AC. Glass optical waveguides: A review of fabrication techniques. Optical Engineering. 2014;**53**(7):071819. DOI: 10.1117/1.OE.53.7.071819

[18] Esposito S. "Traditional" Sol-gel chemistry as a powerful tool for the preparation of supported metal and metal oxide catalysts. Materials. 2019;**12**(4):668. DOI: 10.3390/ma12040668

[19] Biswas PK. Journal of Sol-Gel Science and Technology. 2011;**59**:456. DOI: 10.1007/s10971-010-2368-5

[20] Ho C, Ngo QN, Pita K. Understanding the Cause of Cracking in Sol-Gel-Derived Films. 2007. DOI: 10.13140/RG.2.1.4084.8724

[21] Hill C, Jones S, Boys D. Rapid thermal annealing - theory and practice. In: Levy RA, editor. Reduced Thermal Processing for ULSI. Nato asi Series (Series B: Physics). Vol. 207. Boston, MA: Springer; 1989

[22] Selvaraja SK, Sethi P. Review on optical waveguides. In: You KY, editor. Emerging Waveguide Technology. IntechOpen; 2018. DOI: 10.5772/intechopen.77150. Available from: https://www.intechopen.com/books/emerging-waveguide-technology/review-on-optical-waveguides

[23] Orignac X, Vasconcelos H, Du X, et al. Journal of Sol-Gel Science and Technology. 1997;**8**:243. DOI: 10.1023/A:1026497826255

[24] Guozhong C, Ying W. Nanostructures and Nanomaterials: Synthesis, Properties, and Applications

[25] Langlet M. Low-temperature processing of sol–gel thin films in the SiO_2–TiO_2 binary system. In: Klein L, Aparicio M, Jitianu A, editors. Handbook of sol-Gel Science and Technology. Cham: Springer; 2017. DOI: https://doi.org/10.1007/978-3-319-19454-7_15-1

[26] Marques AC, Almeida RM. Raman spectra and structure of multicomponent oxide planar waveguides prepared by sol-gel. Journal of Sol-Gel Science and Technology. 2006;**40**:371-378. DOI: 10.1007/s10971-006-9320-

[27] Fardad A, Andrews M, Milova G, Malek-Tabrizi A, Najafi I. Fabrication of ridge waveguides: A new solgel route. Applied Optics. 1998;**37**:2429-2434

[28] Mackenzie JD. Journal of Sol-Gel Science and Technology. 1994;**2**:81. DOI: 10.1007/BF00486217

[29] Zhang X, Xue C, Zhang W, Yu L, Wang Q, Que W, et al. Multifunctional TiO_2/ormosils organic-inorganic hybrid films derived by a sol-gel process for photonics and UV nanoimprint applications. Optical Materials Express. 2019;**9**:304-314

[30] Wang B. Optical and surface properties of hybrid TiO_2/ormosil planar waveguide prepared by the sol–gel process. Ceramics International. 2006;**32**:7-12. DOI: 10.1016/j.ceramint.2004.11.013

[31] Desurvire E, Bayart D, Desthieux B, Bigo S. Erbium Doped Fiber Amplifiers: Device and System Developments. Wiley Interscience; 2002. ISBN: 0-471-41903-6

[32] Natarajan S, Ramamurthi A, Selvarajan A, Muthuraman M, Patil K. Amplification of Light in Sol-Gel-Based Nd-Glass Waveguides; 1998

[33] Vasconcelos H, Pinto A. Fluorescence Properties of Rare-Earth-Doped sol-Gel Glasses. 2017. DOI: 10.5772/intechopen.68534

[34] Vasconcelos HC, Pinto AS. Rare-earth activated glasses in integrated optical devices with different geometric shapes: Fibers, planar waveguides and microspheres. In: Photoluminescence: Advances in Research and Applications. Nova Science Publishers, Inc.; 2018. pp. 53-96

[35] Strohhöfer C, Fick J, Vasconcelos HC, Almeida R. Active optical properties of Er-containing crystallites in sol–gel derived glass films. Journal of Non-Crystalline Solids. 1998;

226:182-191. DOI: 10.1016/S0022-3093(98)00365-2

[36] Santos L, Vasconcelos H, Marques M, Almeida R. Active nanocrystals in erbium-doped silica-titania sol-gel films. Materials Science Forum. 2004;**455–456**:545-549. DOI: 10.4028/www.scientific.net/MSF.455-456.545

[37] Almeida R, Vasconcelos H, Gonçalves MC, Santos L. XPS and NEXAFS studies of rare-earth doped amorphous sol–gel films. Journal of Non-Crystalline Solids. 1998;**232**:65-71. DOI: 10.1016/S0022-3093(98)00545-6

[38] Almeida RM, Vasconcelos HC. Rare-earth doped nanocrystals in planar waveguides. In: Fundamentals of Glass Science and Technology. Vaxjo, Sweden: Glafo, the Glass Research Institute; 1997. pp. 110-117

[39] Vasconcelos H. The effect of PO2,5 and AlO1,5 additions on structural changes and crystallization behavior of SiO_2-TiO_2 sol-gel derived glasses and thin films. Journal of Sol-Gel Science and Technology. 2010;**55**. DOI: 10.1007/s10971-010-2223-8

[40] Zimin LG, Gaponenko SV, Lebed VY, Malinovskii IE, Germanenko IN. Nonlinear optical absorption of CuCl and CdS_xSe_{1-x} microcrystallites under quantum confinement. Journal of Luminescence. 1990;**46**(2):101-107. DOI: 10.1016/0022-2313(90)90012-z

[41] Kovalenko M, Heiss W, Shevchenko E, Lee J-S, Schwinghammer H, Alivisatos, et al. SnTe Nanocrystals: A new example of narrow-gap semiconductor quantum dots. Journal of the American Chemical Society. 2007;**129**:11354-11355. DOI: 10.1021/ja074481z

[42] Dawnay EJC, Fardad MA, Green M, Horowitz F, Yeatman EM, Almeida RM, et al. In: Vincenzini P, Righini GC, editors. Advanced Materials in Optics, Electro-Optics and Communication Technologies. Faenza, Italy: Techna; 1995. pp. 55-62

[43] Kajzar F. M-line spectroscopy for nonlinear characterization of polymeric waveguides. Optical Engineering. 1995; **34**(12):3418. DOI: 10.1117/12.213240

[44] Fardad MA, Fick J, Green M, Guntau M, Yeatman EM, Vitrant G, et al. Fabrication and characterisation of a CdS-doped silica-on-silicon planar waveguide. IEE Proceedings - Optoelectronics. 1996;**143**(5):298-302. DOI: 10.1049/ip-opt:19960836

Chapter 4

High FSR and Critical Coupling Control of Microring Resonator Based on Graphene-Silicon Multimode Waveguides

Trung-Thanh Le and Duy-Tien Le

Abstract

We present a new approach for designing a compact microring resonator structure based on only one multimode waveguide, which can provide a very high free spectral range (FSR) and capability of controlling the critical coupling. The silicon on insulator (SOI) waveguide and graphene-silicon waveguide (GSW) are used for the proposed structure. By changing the applied voltage on the graphene sheet, we can achieve a full control of the critical coupling. Some important properties of the proposed microring resonator such as free spectral range and quality factor are analyzed. We show that our structure can provide all characteristics of a single microring resonator with universal applications such as optical switching, modulating, filtering and signal processing, etc.

Keywords: multimode interference (MMI), silicon on insulator, multimode waveguide, directional coupler, finite difference time difference (FDTD), finite difference method (FDM), microring resonators (MRRs), graphene

1. Introduction

In recent years, there has been intense research about ring resonators (RRs) as the building blocks for various photonic applications such as optical switches, wavelength multiplexers, routers, optical delay lines, and optical sensors [1–3]. In the literature, microring resonators with high-quality factors (Q) are required for enhanced nonlinear effects, low threshold lasing, and sensing applications. Almost all of the proposed microring resonator structures used directional couplers as coupling elements. It was shown that such devices are very sensitive to fabrication tolerance [4]. However, a very high Q is undesirable for high-speed signal processing since it can significantly limit the operational bandwidth of the system. In addition, in microring resonator structure, the coupling ratio and loss must be matched so that the ring operates near the critical coupling to achieve a high extinction ratio [5]. To obtain a high bandwidth, a solution is to use a directional coupler with a small gap between two waveguides in order to increase the coupling coefficient. However, this causes large excess mode conversion losses, limiting the flexibility of this approach [6].

Another type of coupler, namely, the 2 × 2 multimode interference (MMI) coupler, has been employed in ring resonators [7–10]. MMI couplers have been shown to have relaxed fabrication requirements and are less sensitive to the wavelength or polarization variations [11, 12]. In recent years, we have presented some microring resonators based on silicon waveguides using 2 × 2, 3 × 3 MMIs for the first time [4, 13–15]. It showed that the proposed devices have good performance compared with structures based on directional couplers.

In a single ring resonator, control of the critical coupling is an important requirement [16–18]. In the literature, the Mach-Zehnder (MZI) configuration is used for this purpose [17]. The key physical mechanism they rely on is the plasma dispersion effect or thermo-optic effect. Another approaches used to create phase shifter are based on silicon-organic hybrid slot waveguides [19] and BTO-Si slot waveguides [20]. The major drawbacks of these schemes are relatively large dimensions. In addition, plasma dispersion can only induce a small variation of the refractive index; the long length of the phase shifter is required. In this chapter, we use the graphene-silicon waveguide (GSW) for phase shifter and controlling the coupling coefficient.

It is noted that in the literature, the MZI configuration for critical coupling control is based on two 3 dB 2 × 2 directional couplers or 2 × 2 MMI coupler. In this chapter, we do not need to use the MZI for controlling the critical coupling of microring resonators, but we present a new way of achieving the critical coupling based on architecture itself. Our approach has advantages of compact size and ease of fabrication with the current CMOS circuit.

In addition, we use the graphene-silicon waveguide for the phase control. Graphene is a single-sheet carbon atom in a hexagonal lattice [21]. Graphene has some potential properties for optical devices. Graphene is a 2-D single-layer carbon atoms arranged in a hexagonal lattice that has raised considerable interest in recent years due to its remarkable optical and electronic properties. For example, graphene has a much higher electron mobility than silicon [22–24]. In particular, it has a linear dispersion relationship in the so-called Dirac points where electrons behave as fermions with zero mass. As a result, we can design optical switches or modulators based on this property. Graphene can also absorb light over a broad-frequency range, so this enables high-speed applications. The density of states of carriers near the Dirac point is low, and the Fermi energy can be tuned significantly with relatively low applied voltage. The Fermi-level tuning changes the refractive index of the graphene. Therefore, the graphene sheet integrated with optical waveguide such as silicon on insulator (SOI) waveguide can provide the possibilities of programmable in optoelectronics.

2. Microring resonator based on multimode waveguides

2.1 Device structure

A new optical microring resonator based on only one multimode waveguide with four ports is shown in **Figure 1**.

Figure 2(a) shows the single-mode waveguide profile. We use silicon on insulator waveguide with a width of 500 nm and height of 250 nm for input and output waveguides. For a multimode waveguide, we use a wider width of $W_{MMI} = 6\mu m$. The field profile of the fundamental mode of the SOI waveguide is shown in **Figure 2(b)**. The refractive indices of silicon and silicon oxide used in our simulations are $n_{Si} = 3.45$, $n_{SiO2} = 1.45$. The field profiles of the fundamental mode and the first-order mode of the multimode waveguide are shown in **Figure 2(c)** and **(d)**.

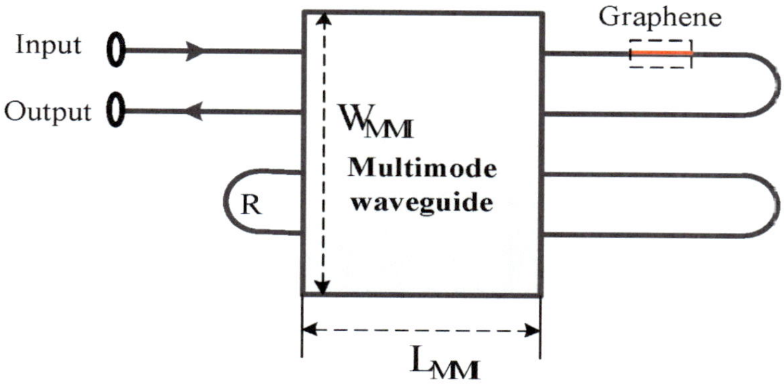

Figure 1.
Microring resonator based on only one multimode waveguide structure.

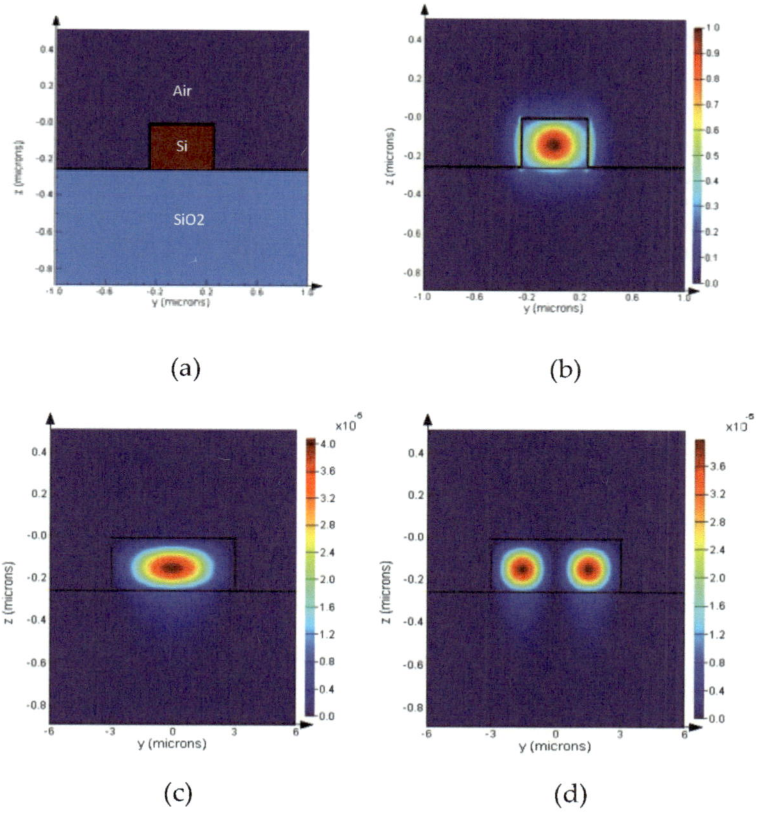

Figure 2.
(a) SOI waveguide structure, (b) field profile of the single-mode SOI waveguide, (c) fundamental mode of the multimode waveguide, and (d) the first-order mode of the multimode waveguide.

In this structure, we use a bent waveguide to connect input port 3 to port 4 as a ring resonator waveguide. Because port 3 is very near to port 4, the bent waveguide radius is relatively small. Therefore, our structure can provide a very high free spectral range (FSR), which is suitable for high-speed communications.

In the next section, we show that our structure can act like a microring resonator. In order to control the critical coupling, we use graphene integrated with silicon waveguide. Graphene can be incorporated into silicon to implement graphene-silicon waveguide. The length of the graphene waveguide is L_{arm}. The cross-section view of the graphene-silicon waveguide is shown in **Figure 3(a)**. The GSW has a monolayer graphene sheet of 340 nm on top of a silicon waveguide, separated from it by a thin Al_2O_3 layer. Graphene, Al_2O_3, and silicon together formed a capacitor structure, which was the basic block of the graphene modulator and phase shifter [25]. The refractive index of Al_2O_3 used in our simulations is 1.6 at the operating wavelength of 1550 nm.

For example, the field profile of the waveguide with a chemical potential $\mu_c = 0.45eV$ is shown in **Figure 3(b)**.

In a multimode waveguide, the information of the image position in the x direction and phases of the output images is very important. We need to know where the multi-images appear in order to design output waveguides to capture the optical output. Furthermore, phase information of the spot images or output images is important for such devices as MMI switch. It can be shown that the field in the multimode region will be of the form [12]

$$f(x, L_{MMI}) = \frac{1}{\sqrt{N}} \sum_{p=0}^{N-1} f_{in}(x - x_p) \exp\left(-j\varphi_p\right) \qquad (1)$$

where $x_p = b(2p - N)\frac{W_{MMI}}{N}$, $\varphi_p = b(N - p)\frac{p\pi}{N}$; $f_{in}(x)$ describes the field profile at the input of the multimode region; x_p and φ_p describe the positions and phases, respectively, of N self-images at that output of the multimode waveguide; p denotes the output image number; and b describes a multiple of the imaging length. For short device, we choose b = 1.

Consider a 4 × 4 multimode waveguide with the length of $L = L_{MMI} = \frac{3L_\pi}{2}$, where $L_\pi = \frac{\pi}{\beta_0 - \beta_1}$ is the beat length of the MMI and β_0, β_1 are the propagation constants of the fundamental and first-order modes supported by the multimode waveguide with a width of W_{MMI}. The phases associated with the images from input i to output j can be presented by

$$\varphi_{ij} = -\frac{\pi}{2}(-1)^{i+j+4} + \frac{\pi}{16}\left[i + j - i^2 - j^2 + (-1)^{i+j+4}\left(2ij - i - j + \frac{1}{2}\right)\right] \qquad (2)$$

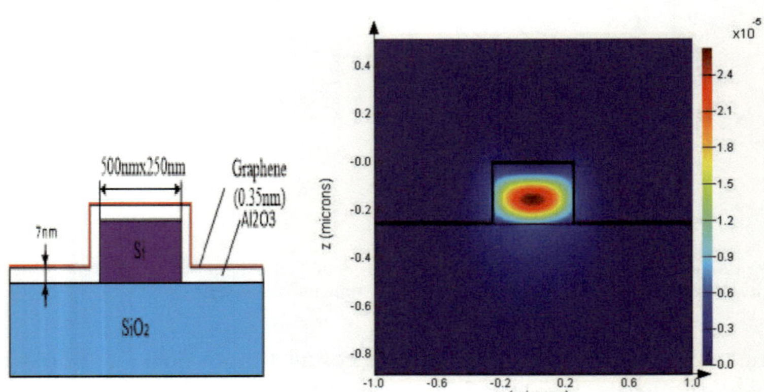

Figure 3.
(a) Graphene-silicon waveguide structure and (b) field profile with a chemical potential of $\mu_c = 0.45eV$.

We showed that the characteristics of an MMI device can be described by a transfer matrix [2]. This transfer matrix is a very useful tool for analyzing cascaded MMI structures. Phase ϕ_{ij} is associated with imaging an input i to an output j in an MMI coupler. These phases ϕ_{ij} form a matrix S_{4x4}, with i representing the row number and j representing the column number. A single 4 × 4 MMI coupler at a length of $L_{MMI} = \frac{3L_\pi}{2}$ is described by the following transfer matrix [26, 27]:

$$\mathbf{S}_{4x4} = \frac{1}{2}\begin{bmatrix} 1-j & 0 & 0 & 1+j \\ 0 & 1-j & 1+j & 0 \\ 0 & 1+j & 1-j & 0 \\ 1+j & 0 & 0 & 1-j \end{bmatrix} \quad (3)$$

The output and input amplitudes at four ports of the 4 × 4 multimode waveguide can be expressed by

$$\overrightarrow{E_{out}} = \begin{pmatrix} E_{out,1} \\ E_{out,2} \\ E_{out,3} \\ E_{out,4} \end{pmatrix} = \mathbf{S}_{4x4}\begin{pmatrix} E_{\text{in},1} \\ E_{\text{in},2} \\ E_{\text{in},3} \\ E_{\text{in},4} \end{pmatrix} = \mathbf{S}_{4x4}\overrightarrow{E_{in}} \quad (4)$$

where $E_{in,i}$ (i = 1, 2, 3, 4) and $E_{\text{out},j}$ (j = 1, 2, 3, 4) are complex amplitudes at input ports and output ports 1–4, respectively. From Eqs. (3) and (4), we can calculate the relations ships between the input and output amplitudes of **Figure 1** as follow:

$$\begin{pmatrix} E_{\text{in},1} \\ E_{\text{in},4} \end{pmatrix} = \mathbf{M}\begin{pmatrix} E_{\text{in},2} \\ E_{\text{in},3} \end{pmatrix} = e^{j\frac{\Delta\varphi}{2}}\begin{bmatrix} \tau & \kappa \\ \kappa^* & -\tau^* \end{bmatrix}\begin{pmatrix} E_{\text{in},2} \\ E_{\text{in},3} \end{pmatrix} \quad (5)$$

where $\tau = \cos\left(\frac{\Delta\varphi}{2}\right)$, and $\kappa = \sin\left(\frac{\Delta\varphi}{2}\right)$, $\Delta\varphi$is the phase difference between the graphene-silicon waveguide with the length of L_{arm} and the silicon on insulator waveguide and can be calculated by [28]

$$\Delta\varphi = \frac{2\pi}{\lambda}\Delta n_{eff}L_{arm} \quad (6)$$

The phase difference $\Delta\varphi$ can be controlled by applying a voltage V_g to the graphene sheet. The field propagation of the multimode waveguide for input ports 1 and 2 is shown in **Figure 4**. The optimal length of the MMI is calculated by the 3D-BPM [29]. We show that the optimal length is found to be 214 μm.

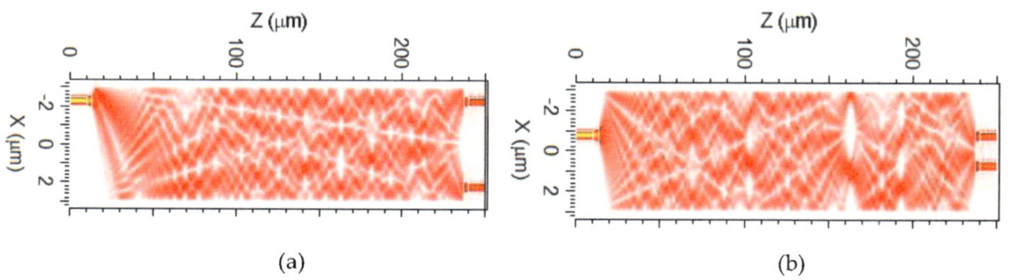

Figure 4.
Field propagation of 4 × 4 MMI coupler: (a) field propagation, in 1, and (b) field propagation, in 2.

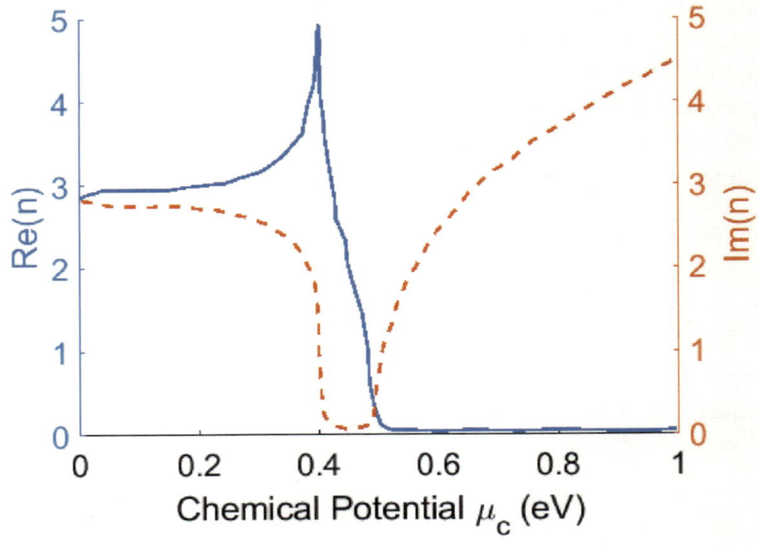

Figure 5.
Refractive index of graphene sheet.

The light propagation through the resonator is characterized by a round-trip transmission $E_{in,3} = \alpha \exp(j\theta) E_{in,4}$, where $\theta = \frac{2\pi}{\lambda} n_{eff} L_R$ is the round-trip phase, α is the loss factor, n_{eff} is the effective refractive index of the SOI single-mode waveguide, and L_R is the ring resonator circumference. The normalized transmitted power of the device can be calculated by

$$T = \left|\frac{E_{in,2}}{E_{in,1}}\right|^2 = \frac{\alpha^2 - \cos^2(\Delta\varphi/2) - 2\alpha|\cos(\Delta\varphi/2)|\cos(\theta)}{1 + \alpha^2\cos^2(\Delta\varphi/2) - 2\alpha|\cos(\Delta\varphi/2)|\cos(\theta)} \tag{7}$$

At resonance wavelengths when $\theta = 2m\pi$, m $= 1, 2, 3, \ldots$, the normalized power transmission is

$$T = \left|\frac{E_{in,2}}{E_{in,1}}\right|^2 = \frac{|\alpha - |\cos(\Delta\varphi/2)||^2}{|1 - \alpha|\cos(\Delta\varphi/2)||^2} \tag{8}$$

2.2 Graphene-silicon waveguide

The presence of the graphene layer changes the propagation characteristics of the guided modes, and these can be controlled and reconfigured, changing the chemical potential by means of applying a suitable voltage V_g. The real and image parts of the refractive index of graphene with different chemical potentials are shown in **Figure 5** [30].

Graphene has optical properties due to its band structure that provides both intraband and interband transitions. Both types of the transitions contribute to the material conductivity expressed by

$$\sigma(\omega) = \sigma_{\text{intra}}(\omega) + \sigma_{\text{inter}}(\omega) \tag{9}$$

where $\sigma_{\text{intra}}(\omega)$ and $\sigma_{\text{inter}}(\omega)$ are the intraband and interband conductivities, which can be calculated by the Kubo's theory:

$$\sigma_{\text{intra}}(\omega) = \frac{ie^2}{\pi\hbar(\omega + i2\Gamma)}\left[\frac{\mu_c}{k_B T} + 2\ln\left(e^{-(\mu_c/k_B T)} + 1\right)\right] \quad (10)$$

$$\sigma_{\text{inter}}(\omega) = -\frac{ie^2}{4\pi\hbar}\ln\left(\frac{2|\mu_c| - (\omega - 2i\Gamma)\hbar}{2|\mu_c| + (\omega - 2i\Gamma)\hbar}\right) \quad (11)$$

where e is electron charge, $\hbar$ is the angular Planck constant, k_B is the Boltzmann constant, T is the temperature, μ_c is the Fermi level or chemical potential, $\Gamma = \frac{eV_F^2}{\mu\mu_c}$ is the electron collision rate, μ is electron mobility, and V_F is the Fermi velocity in graphene.

The dielectric constant of a graphene layer can be calculated by [21, 22]

$$\varepsilon_g(\omega) = 1 + \frac{i\sigma(\omega)}{\omega\varepsilon_0\Delta} \quad (12)$$

The refractive index of the graphene layer sheet can be changed by providing the applied voltage V_g to the graphene sheet. It is because it will change the value of the chemical potential:

$$\left|\mu_c(V_g)\right| = \hbar V_F\sqrt{\pi\left|\eta(V_g - V_0)\right|} \quad (13)$$

where V_0 is the offset voltage from zero caused by natural doping.

2.3 Critical coupling control

It is shown that the normalized transmission T through the device can be switched from unity to zero at the condition of critical coupling, given by $\alpha = |\tau|$. The control of the phase shift $\Delta\varphi$, so the condition of the critical coupling is met, can be achieved through applying voltage to the graphene sheet. The effective index of the graphene-silicon waveguide calculated at different chemical potentials by FDM method is shown in **Figure 6**. We see that for low-loss waveguide, the

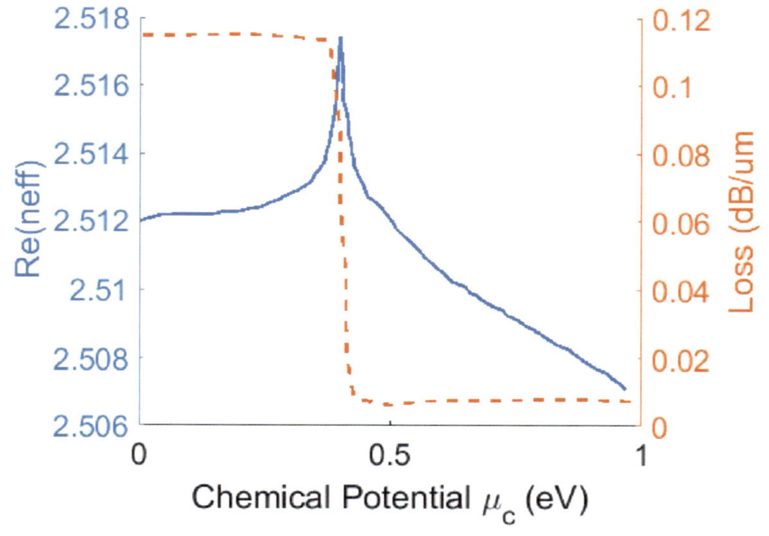

Figure 6.
Effective refractive index of the GSW waveguide.

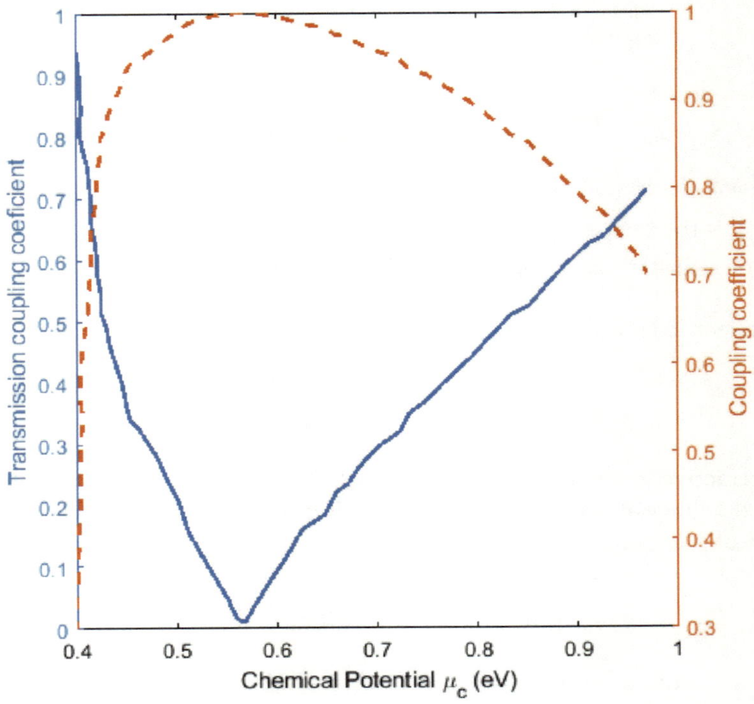

Figure 7.
Transmission and coupling coefficients of the resonator with different chemical potentials.

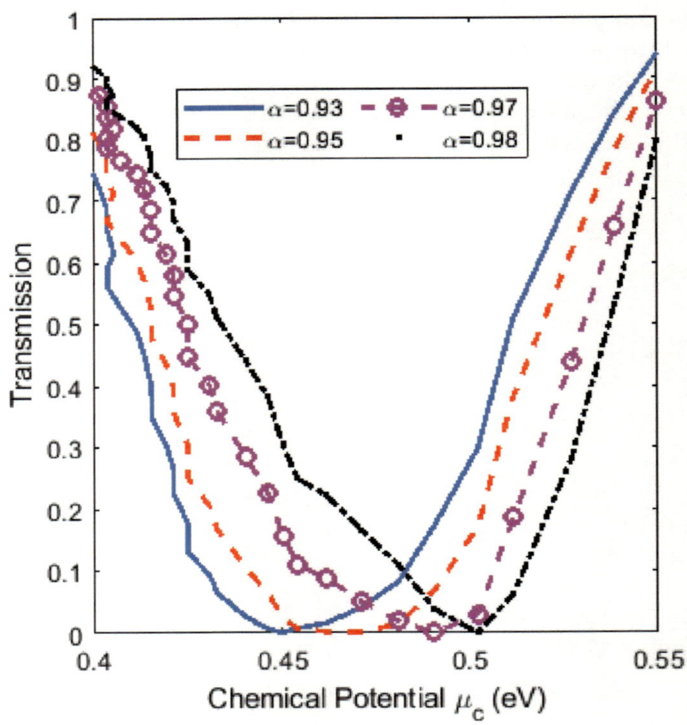

Figure 8.
Transmissions of the microring resonator with different chemical potential and loss factor.

chemical potential should be larger than $\mu_c = 0.4eV$. By changing the chemical potential, the transmission and coupling coefficients $|\tau|, |\kappa|$ can be changed as shown in **Figure 7**. The simulations show that we can get full control of the coupling coefficient from zero to unity by changing the chemical potential from $\mu_c = 0.4eV$ to $\mu_c = 0.58eV$.

The transmissions of the device at different chemical potentials and loss factors are shown in **Figure 8**. The simulations have two very important features which are the key for most of the proposed applications: (1) The transmitted power is zero at a value of critical coupling, and (2) for high-quality factor, the portion of the curve to the right of the critical coupling point is steep. Small changes of the phase shifter can control the transmitted power and switch between unity and zero [17]. This chapter shows that we can achieve high-speed devices based on our proposed microring resonator.

Some other performance parameters of the microring resonator are finesse, Q-factor, resonance width, and bandwidth. These are all terms that are mainly related to the full width at half of the maximum (FWHM) of the transmission. The quality factor Q of the microring resonator of the structure in **Figure 1** can be derived as [3]

$$Q = \frac{\pi N_g L_R}{\lambda} \frac{\sqrt{\alpha\tau}}{1 - \alpha\tau} \tag{14}$$

Another important parameter for microring resonators is the finesse F, which is defined and calculated for the single- and add-drop microring resonators by

$$F = \frac{FSR}{\Delta\lambda_{FWHM}} = \pi \frac{\sqrt{\alpha\tau}}{1 - \alpha\tau} \tag{15}$$

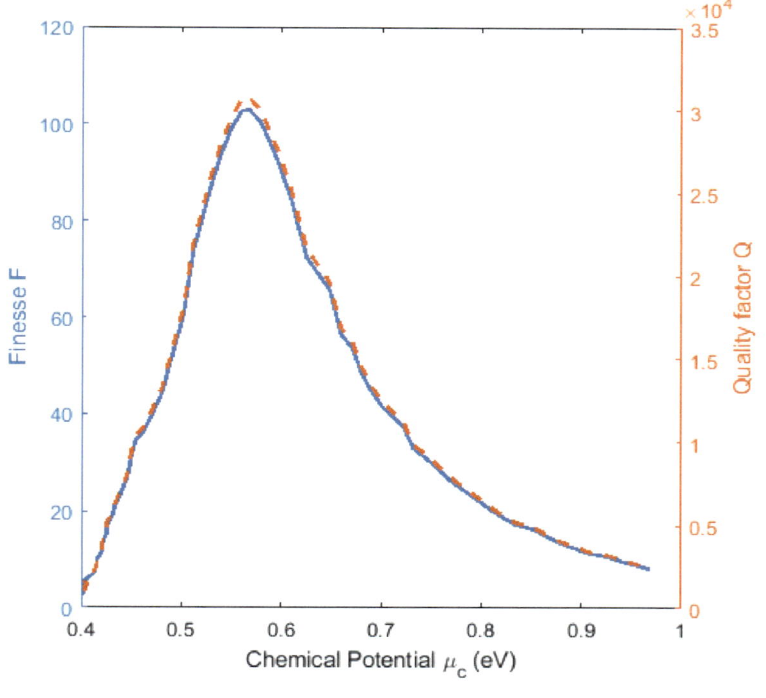

Figure 9.
Finesse and quality factor at different chemical potentials.

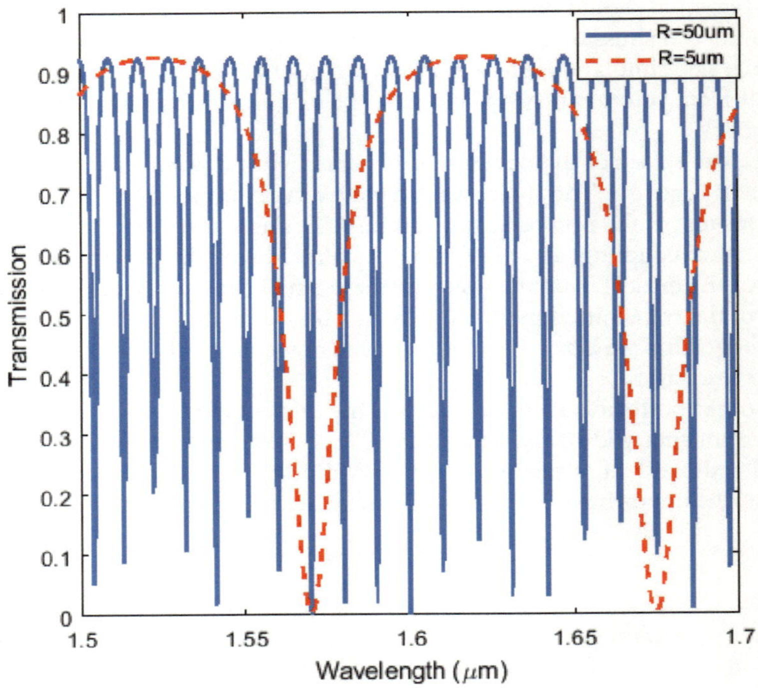

Figure 10.
Transmissions of the microring resonator with two microring radii of 5μm and 50μm.

where $\Delta\lambda_{FWHM}$ is the resonance full width at half maximum and FSR is the free spectral range. The free spectral range is the distance between two peaks on a wavelength scale. By differentiating the equation $\varphi = \beta L_R$, we get $FSR = \frac{\lambda^2}{n_g L}$, where the group index $n_g = n_{eff} - \lambda \frac{dn_{eff}}{d\lambda}$.

Figure 9 shows the finesse and quality factor with different chemical potential at a radius of $5\mu m$. We see that a maximum finesse and quality factor can be achieved at a chemical potential of $\mu_c = 0.57eV$.

The normalized transmissions of the propose microring resonator in **Figure 1** at microring radii of $5\mu m$ and $50\mu m$ are shown in **Figure 10**. Here we assume that the chemical potential is $\mu_c = 0.45eV$. The simulations show that the exact characteristics of a single microring resonator can be achieved.

Finally, we use FDTD method to simulate the proposed microring resonator based on multimode waveguide. In our FDTD simulations, we take into account the wavelength dispersion of the silicon waveguide. A light pulse of 15 fs pulse width is launched from the input to investigate the transmission characteristics of the device. The grid sizes Δx = Δy = Δz = 20 nm are chosen in our simulations for

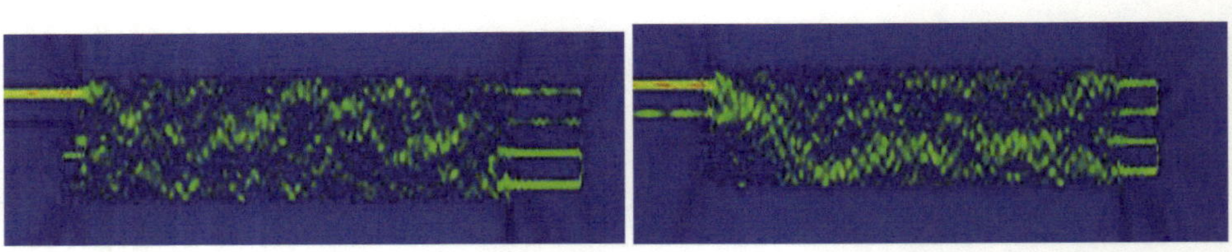

(a) (b)

Figure 11.
Optical field propagation through the coupler for input signal presented at (a) port 1 and (b) port 2.

accurate simulations [31]. The FDTD simulations for the proposed microring resonator with chemical potential of 0.45 and 0.42 eV are shown in **Figure 11(a)** and **(b)**. The simulations show that the device operation has a good agreement with our prediction by analytical theory.

3. Conclusions

We presented a new microring resonator based on only one multimode waveguide. The critical control of the microring resonator can be achieved by using graphene-silicon waveguide. The proposed device has all characteristics of a traditional microring resonator. Some important parameters of the proposed device such as finesse, quality factor, etc. are also presented in this chapter. The device operation has been verified by using the FDTD. This microring resonator structure is very compact and can be useful for further applications in optical switching, filtering, and sensing.

Acknowledgement

This research is funded by the Ministry of Natural Resources and Environment of Vietnam under the project BĐKH.30/16-20.

Author details

Trung-Thanh Le* and Duy-Tien Le
International School (VNU-IS), Vietnam National University (VNU), Hanoi, Vietnam

*Address all correspondence to: thanh.le@vnu.edu.vn

References

[1] Le T-T. Two-channel highly sensitive sensors based on 4 × 4 multimode interference couplers. Photonic Sensors. 2017;**20**:1-8. DOI: 10.1007/s13320-017-0441-1

[2] Le T-T. Multimode Interference Structures for Photonic Signal Processing. Denmark: LAP Lambert Academic Publishing; 2010

[3] Chremmos I, Schwelb O, Uzunoglu N, editors. Photonic Microresonator Research and Applications. New York: Springer; 2010

[4] Le D-T, Le T-T. Fano resonance and EIT-like effect based on 4x4 multimode interference structures. International Journal of Applied Engineering Research. 2017;**12**(13):3784-3788

[5] Yariv A. Universal relations for coupling of optical power between microresonators and dielectric waveguides. Electronics Letters. 2000; **36**:321-322

[6] Vlasov Y, McNab S. Losses in single-mode silicon-on-insulator strip waveguides and bends. Optics Express. 2004;**12**:1622-1631

[7] Xu DX, Densmore A, Waldron P, Lapointe J, Post E, Delâge A. High bandwidth SOI photonic wire ring resonators using MMI coupler. Optics Express. 2007;**15**:3149-3155

[8] Cahill L, Le T. The design of signal processing devices employing SOI MMI couplers. In: Proceedings of the SPIE of Paper 7220–2, Integrated Optoelectronic Devices (OPTO 2009), Photonics West. San Jose, California, USA: San Jose Convention Center; 2009. pp. 24-29

[9] Cahill LW, Le TT. MMI devices for photonic signal processing. In: 9th International Conference on Transparent Optical Networks (ICTON 2007). Rome, Italy; 2007. pp. 202-205

[10] Le TT, Cahill LW. Microresonators based on 3x3 restricted interference MMI couplers on an SOI platform. In: The IEEE LEOS Conference, 2009 (LEOS '09), Belek-Antalya. Turkey; 2009. pp. 479-480

[11] Soldano LB. Multimode interference couplers [PhD Thesis]. Delft, the Netherlands: Delft University of Technology; 1994

[12] Bachmann M, Besse PA, Melchior H. General self-imaging properties in N x N multimode interference couplers including phase relations. Applied Optics. 1994;**33**(18):3905-3911

[13] Le T-T. Microring resonator based on 3x3 general multimode interference structures using silicon waveguides for highly sensitive sensing and optical communication applications. International Journal of Applied Sciences and Engineering. 2013;**11**: 31-39

[14] Le T-T, Cahill L. Generation of two Fano resonances using 4x4 multimode interference structures on silicon waveguides. Optics Communications. 2013;**301-302**:100-105

[15] Le D-T, Do T-D, Nguyen V-K, Nguyen A-T, Le T-T. Sharp asymmetric resonance based on 4x4 multimode interference coupler. International Journal of Applied Engineering Research. 2017;**12**(10):2239-2242

[16] Choi JM, Lee RK, Yariv A. Control of critical coupling in a ring resonator–fiber configuration: Application to wavelength-selective switching, modulation, amplification, and oscillation. Optics Letters. 2001;**26**: 1236-1238

[17] Yariv A. Critical coupling and its control in optical waveguide-ring resonator systems. IEEE Photonics Technology Letters. 2002;**14**:483-485

[18] Le T-T. Control of critical coupling in 3x3 mmi couplers based on optical microring resonators and applications to selective wavelength switching, modulation, amplification and oscillation. Journal of Science and Technology. 2017;**6**:24-28

[19] Steglich P, Mai C, Villringer C, Pulwer S, Casalboni M, Schrader S, et al. Quadratic electro-optic effect in silicon-organic hybrid slot-waveguides. Optics Letters. 2018;**43**:3598-3601

[20] Abel S, Eltes F, Ortmann JE, Messner A, Castera P, Wagner T, et al. Large Pockels effect in micro- and nanostructured barium titanate integrated on silicon. Nature Materials. 2019;**18**:42-47

[21] Hanson GW. Dyadic Green's functions and guided surface waves for a surface conductivity model of graphene. Journal of Applied Physics. 2008;**103**:064302

[22] Capmany J, Domenech D, Muñoz P. Silicon graphene Bragg gratings. Optics Express. 2014;**22**:5283-5290

[23] Midrio M, Galli P, Romagnoli M, Kimerling LC, Michel J. Graphene-based optical phase modulation of waveguide transverse electric modes. Photonics Research. 2014;**2**:A34-A40

[24] Xing P, Ooi KJA, Tan DTH. Ultra-broadband and compact graphene-on-silicon integrated waveguide mode filters. Scientific Reports. 2018;**8**:9874

[25] Bao Q. 2D Materials for Photonic and Optoelectronic Applications. United Kingdom: Woodhead Publishing; 2019

[26] Le T-T. Two-channel highly sensitive sensors based on 4 × 4 multimode interference couplers. Photonic Sensors. 2017;7:357-364

[27] Le T-T, Cahill L. The design of 4×4 multimode interference coupler based microring resonators on an SOI platform. Journal of Telecommunications and Information Technology. 2009;**2**:98-102

[28] Petrone G, Cammarata G. Modelling and Simulation. UK: InTech Publisher; 2008

[29] Le T-T. An improved effective index method for planar multimode waveguide design on an silicon-on-insulator (SOI) platform. Optica Applicata. 2013;**43**:271-277

[30] Amin R, Ma Z, Maiti R, Khan S, Khurgin JB, Dalir H, et al. Attojoule-efficient graphene optical modulators. Applied Optics. 2018;**57**:D130-D140

[31] Rumley S, Bahadori M, Polster R, Hammond SD, Calhoun DM, Wen K, et al. Optical interconnects for extreme scale computing systems. Parallel Computing. 2017;**64**:65-80

Section 3

Waveguides in Microwave Engineering

Chapter 5

Rigorous Analysis of the Propagation in Metallic Circular Waveguide with Discontinuities Filled with Anisotropic Metamaterial

Hedi Sakli and Wyssem Fathallah

Abstract

In this chapter, we present an extension of the rigorous analysis of the propagation of electromagnetic waves in magnetic transverse (TM) and transverse electric (TE) modes in a metallic circular waveguide partially filled with anisotropic metamaterial. In our analysis, the design of waveguide filters with uniaxial discontinuities is based on the determination of the higher-order modes, which have been analyzed and exploited. Below the cutoff frequency, the back backward waves can propagate in an anisotropic material. The numerical results with our MATLAB code for TM and TE modes were compared to theoretical predictions. Good agreements have been obtained. We analyzed a waveguide filters filled with partially anisotropic metamaterial using the mode matching (MM) technique based on the Scattering Matrix Approach (SMA), which, from the decomposition of the modal fields (TE and TM modes), are used to determine the dispersion matrix and thus the characterization of a discontinuity in waveguide. We extended the application of MM technique to the anisotropic material. By using modal analysis, our approach has considerably reduced the computation time compared to High Frequency Structure Simulator (HFSS) software.

Keywords: anisotropic metamaterials, forward and backward waves, MM, modal analysis, waveguides discontinuity

1. Introduction

Guided modes in circular waveguides consist of metamaterials [1–13] have been studied in the literature. Many studies of propagation modes in this waveguides with isotropic media [14–17] or double negative metamaterials [18, 19] have been presented in the literature. However, the rigorous study of the dispersion of anisotropic metamaterials in circular waveguides presents a lack in the literature. In this chapter, we present an extension of the rigorous analysis of the propagation of electromagnetic waves in magnetic transverse (TM) and electric transverse (TE) modes in the case of anisotropic circular waveguides, who take account of the spatial distribution of the permittivity and permeability of the medium. In this structure, the propagation modes are exploited. The effects of anisotropic parameter on cutoff frequencies and dispersion characteristics are discussed. Below the cutoff frequency, the back backward waves can propagate in an anisotropic material. The numerical results with our MATLAB code for TM and TE modes were compared to theoretical predictions. Good agreements have been obtained. We analyzed a waveguide filters filled with partially anisotropic metamaterial using the mode matching (MM) technique based on the Scattering Matrix Approach (SMA) which, from the decomposition of the modal fields, are used to determine the dispersion matrix and thus the characterization of a discontinuity in waveguide. We extended the application of MM technique to the anisotropic material.

This formulation can be a useful tool for engineers of microwave. The metamaterial is largely applied by information technology industries, particularly in the radio frequency devices and microwaves such as the waveguide antennas, the patch antennas, the circulators, the resonators and the filters.

2. Formulation

In the anisotropic diagonal metamaterials medium, the Maxwell equations are expressed as follows

$$\vec{\nabla} \times \vec{E} = -j\omega\overline{\overline{\mu}}.\vec{H} \tag{1}$$

$$\vec{\nabla} \times \vec{H} = j\omega\overline{\overline{\varepsilon}}.\vec{E} \tag{2}$$

with

$$\overline{\overline{\mu}} = \mu_0 \begin{pmatrix} \mu_{rr} & 0 & 0 \\ 0 & \mu_{r\theta} & 0 \\ 0 & 0 & \mu_{rz} \end{pmatrix} = \mu_0 \begin{pmatrix} \mu_{rt} & 0 \\ 0 & \mu_{rz} \end{pmatrix}. \tag{3}$$

and

$$\overline{\overline{\varepsilon}} = \varepsilon_0 \begin{pmatrix} \varepsilon_{rr} & 0 & 0 \\ 0 & \varepsilon_{r\theta} & 0 \\ 0 & 0 & \varepsilon_{rz} \end{pmatrix} = \varepsilon_0 \begin{pmatrix} \varepsilon_{rt} & 0 \\ 0 & \varepsilon_{rz} \end{pmatrix} \tag{4}$$

Let consider a circular waveguide of radius R completely filled with anisotropic metamaterial without losses, as represented in the **Figure 1**. The wall of the guide is perfect conductor.

By considering the propagation in the Oz direction and manipulating Eqs. (1) and (2), we obtain the expressions of the transverse electromagnetic fields according to the longitudinal fields.

$$E_r = \frac{-j}{K_{c.r}^2}\left(k_z \frac{\partial E_z}{\partial r} + \frac{\omega\mu_0\mu_{r\theta}}{r}\frac{\partial H_z}{\partial \theta}\right) \tag{5}$$

$$E_\theta = \frac{j}{K_{c.\theta}^2}\left(\frac{-k_z}{r}\frac{\partial E_z}{\partial \theta} + \omega\mu_0\mu_{rr}\frac{\partial H_z}{\partial r}\right) \tag{6}$$

$$H_r = \frac{-j}{K_{c.\theta}^2}\left(-\frac{\omega\varepsilon_0\varepsilon_{r\theta}}{r}\frac{\partial E_z}{\partial \theta} + k_z\frac{\partial H_z}{\partial r}\right) \tag{7}$$

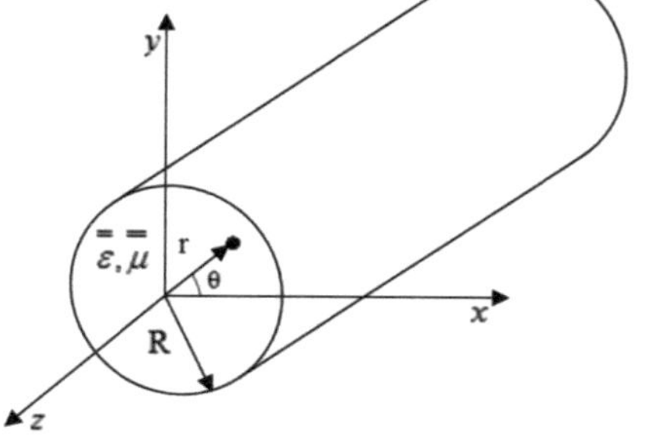

Figure 1.
Geometry of circular waveguide filled with metamaterial.

$$H_\theta = \frac{j}{K_{c.r}^2}\left(-\omega\varepsilon_0\varepsilon_{rr}\frac{\partial E_z}{\partial r} - \frac{k_z}{r}\frac{\partial H_z}{\partial \theta}\right) \tag{8}$$

with

$$K_{c.r}^2 = k_0^2\varepsilon_{rr}\mu_{r\theta} - k_z^2 \tag{9}$$

$$K_{c.\theta}^2 = k_0^2\varepsilon_{r\theta}\mu_{rr} - k_z^2 \tag{10}$$

$$k_0^2 = \omega^2\varepsilon_0\mu_0 \tag{11}$$

When E and H are the electric and magnetic field respectively. ε and μ are the permittivity and permeability. k_z is the propagation constant in z-direction.

In this chapter, we study rigorously the TE and TM modes in this anisotropic waveguide.

2.1 Transverse electric (TE) modes

From Eq. (1), the differential equation for z-component can be obtained as follows

$$\frac{\partial^2 H_z}{\partial r^2} + \frac{1}{r}\frac{\partial H_z}{\partial r} + \left(\frac{K_{c.\theta}^{(h)}\cdot\sqrt{\mu_{r\theta}}}{K_{c.r}^{(h)}\cdot\sqrt{\mu_{rr}}}\right)^2\frac{1}{r^2}\frac{\partial^2 H_z}{\partial \theta^2} + \left(\frac{\sqrt{\mu_{rz}}}{\sqrt{\mu_{rr}}}K_{c.\theta}^{(h)}\right)^2 H_z = 0. \tag{12}$$

Using the separation of the variables (r,θ), the expression of the longitudinal magnetic field H_z for the TE_{mn} modes in the circular metallic waveguide completely filled with anisotropic metamaterial is necessary for the resolution of the differential Eq. (12). H_z can be written as follows

$$H_z^{(h)} = H_0 \sin\left(\frac{K_{c.\theta}^{(h)}\cdot\sqrt{\mu_{r\theta}}}{K_{c.r}^{(h)}\cdot\sqrt{\mu_{rr}}}n.\theta\right)J_n\left(\frac{\sqrt{\mu_{rz}}}{\sqrt{\mu_{rr}}}K_{c.\theta}^{(h)}.r\right)e^{-jk_z z} \tag{13}$$

J_n is the Bessel function of the first kind of order n (n = 0, 1, 2, 3, ...).

The expressions (5)–(8) become

$$E_r^{(h)} = \frac{-j\omega\mu_0\mu_{r\theta}}{K_{c.r}^2.r}\left(\frac{K_{c.\theta}^{(h)}\cdot\sqrt{\mu_{r\theta}}}{K_{c.r}^{(h)}\cdot\sqrt{\mu_{rr}}}n.\theta\right)H_0.\cos\left(\frac{K_{c.\theta}^{(h)}\cdot\sqrt{\mu_{r\theta}}}{K_{c.r}^{(h)}\cdot\sqrt{\mu_{rr}}}n.\theta\right)J_n\left(\frac{\sqrt{\mu_{rz}}}{\sqrt{\mu_{rr}}}K_{c.\theta}^{(h)}.r\right)e^{-jk_z z} \tag{14}$$

$$E_\theta^{(h)} = \frac{j\omega\mu_0\mu_{rr}}{K_{c.\theta}^2}\left(\frac{\sqrt{\mu_{rz}}}{\sqrt{\mu_{rr}}}K_{c.\theta}^{(h)}\right)H_0.\sin\left(\frac{K_{c.\theta}^{(h)}\cdot\sqrt{\mu_{r\theta}}}{K_{c.r}^{(h)}\cdot\sqrt{\mu_{rr}}}n.\theta\right)J_n'\left(\frac{\sqrt{\mu_{rz}}}{\sqrt{\mu_{rr}}}K_{c.\theta}^{(h)}.r\right)e^{-jk_z z} \quad (15)$$

$$H_r^{(h)} = \frac{-jk_z}{K_{c.\theta}^2}\left(\frac{\sqrt{\mu_{rz}}}{\sqrt{\mu_{rr}}}K_{c.\theta}^{(h)}\right)H_0.\sin\left(\frac{K_{c.\theta}^{(h)}\cdot\sqrt{\mu_{r\theta}}}{K_{c.r}^{(h)}\cdot\sqrt{\mu_{rr}}}n.\theta\right)J_n'\left(\frac{\sqrt{\mu_{rz}}}{\sqrt{\mu_{rr}}}K_{c.\theta}^{(h)}.r\right)e^{-jk_z z} \quad (16)$$

$$H_\theta^{(h)} = \frac{-jk_z}{K_{c.r}^2.r}\left(\frac{K_{c.\theta}^{(h)}\cdot\sqrt{\mu_{r\theta}}}{K_{c.r}^{(h)}\cdot\sqrt{\mu_{rr}}}n\right)H_0.\cos\left(\frac{K_{c.\theta}^{(h)}\cdot\sqrt{\mu_{r\theta}}}{K_{c.r}^{(h)}\cdot\sqrt{\mu_{rr}}}n.\theta\right)J_n\left(\frac{\sqrt{\mu_{rz}}}{\sqrt{\mu_{rr}}}K_{c.\theta}^{(h)}.r\right)e^{-jk_z z} \quad (17)$$

With J_n' is the derivative of the Bessel function of the first kind of order n (n = 0, 1, 2, 3, ...).

The boundary conditions are written as follows:

$$E_\theta(r = R) = E_z(r = R) = 0 \quad (18)$$

Consequently, from Eq. (15), we obtain

$$J_n'\left(\frac{\sqrt{\mu_{rz}}}{\sqrt{\mu_{rr}}}K_{c.\theta}^{(h)}.R\right) = 0 \quad (19)$$

This implies

$$u_{nm}' = \frac{\sqrt{\mu_{rz}}}{\sqrt{\mu_{rr}}}K_{c.\theta}^{(h)}.R \quad (20)$$

Where u_{nm}' represents the mth zero (m = 1, 2, 3, ...) of the derivative of the Bessel function J_n' of the first kind of order n.

The constant H_0 is determined by normalizing the power flow down the circular guide.

$$P^{TE} = \int_0^R\int_0^{2\pi}\left(E_r^{(h)}H_\theta^{*(h)} - E_\theta^{(h)}H_r^{*(h)}\right)rdrd\theta = 1 \quad (21)$$

Where * indicates the complex conjugate.

Eq. (21) gives

$$H_0 = \frac{K_{c.r}^3}{\sqrt{\omega\mu_0 k_z}}\frac{\sqrt{\mu_{rz}}}{\mu_{r\theta}}N_{nm}^{(h)} \quad (22)$$

With

$$N_{nm}^{(h)} = \frac{1}{\sqrt{\frac{\sigma_n}{2}}\cdot\left(\left(u_{nm}'\right)^2 - n^2\right)^{1/2}\cdot J_n\left(u_{nm}'\right)} \quad (23)$$

$$\sigma_n = \begin{cases} 2\pi, & \text{if } n = 0 \\ \pi - \dfrac{\sin(4\pi a.\text{n})}{4\text{a.n}}, & \text{if } n > 0 \end{cases} \quad (24)$$

$$a = \frac{K_{c.\theta}^{(h)}\cdot\sqrt{\mu_{r\theta}}}{K_{c.r}^{(h)}\cdot\sqrt{\mu_{rr}}} \quad (25)$$

Finally, the propagation constant in TE mode is given by

$$k_{z.nm}^{(TE)} = \pm\sqrt{k_0^2\varepsilon_{r\theta}.\mu_{rr} - \frac{\mu_{rr}}{\mu_{rz}}\left(\frac{u'_{nm}}{R}\right)^2} \tag{26}$$

The cutoff frequency is written

$$f_{c.nm}^{(TE)} = \frac{c}{2\pi}\frac{1}{\sqrt{|\varepsilon_{r\theta}\mu_{rz}|}}\cdot\left(\frac{u'_{nm}}{R}\right). \tag{27}$$

We can introduce the following effective permeability and effective permittivity to describe the propagation characteristics of the waveguide modes [6, 7, 13].

$$\mu_{r,eff}^{TE} = \mu_{rr}, \tag{28}$$

$$\varepsilon_{r,eff}^{TE} = \varepsilon_{r\theta}\left(1 - \frac{1}{\varepsilon_{r\theta}\mu_{rz}k_0^2}\cdot\left(\frac{u'_{nm}}{R}\right)^2\right). \tag{29}$$

Further, it is apparent that:

- $k_z^{TE} = k_0\sqrt{\mu_{r,eff}^{TE}.\varepsilon_{r,eff}^{TE}} > 0$, for $\mu_{r,eff}^{TE} > 0$ and $\varepsilon_{r,eff}^{TE} > 0$;
- $k_z^{TE} = -k_0\sqrt{\mu_{r,eff}^{TE}.\varepsilon_{r,eff}^{TE}} < 0$, for $\mu_{r,eff}^{TE} < 0$ and $\varepsilon_{r,eff}^{TE} < 0$;
- $k_z^{TE} = \pm jk_0\sqrt{\mu_{r,eff}^{TE}.\varepsilon_{r,eff}^{TE}}$, for $\mu_{r,eff}^{TE}.\varepsilon_{r,eff}^{TE} < 0$.

The sign of $\varepsilon_{r,eff}^{TE}$ depends on the sign of μ_{rz}. In the following, we will consider all cases that arise from the different sign of μ_{rz}.

2.1.1 First case $\mu_{rz} > 0$

For $\varepsilon_{r\theta} > 0$, we have.

$$\varepsilon_{r,eff}^{TE} = |\varepsilon_{r\theta}|\left(1 - \frac{1}{|\varepsilon_{r\theta}\,\mu_{rz}|k_0^2}\cdot\left(\frac{u'_{nm}}{R}\right)^2\right) = |\varepsilon_{r\theta}|\left(1 - \left(\frac{f_{c.nm}^{TE}}{f}\right)^2\right) < 0,\ \text{if}\ f < f_{c.nm}^{TE} \tag{30}$$

And for $\varepsilon_{r\theta} < 0$, $\varepsilon_{r,eff}^{TE}$ is rewritten as

$$\varepsilon_{r,eff}^{TE} = -|\varepsilon_{r\theta}|\left(1 + \frac{1}{|\varepsilon_{r\theta}\,\mu_{rz}|k_0^2}\cdot\left(\frac{u'_{nm}}{R}\right)^2\right) < 0. \tag{31}$$

It can be seen that $\mu_{rz} > 0$ leads to $\varepsilon_{r,eff}^{TE} < 0$ below the cutoff frequency whenever $\varepsilon_{r\theta} > 0$ or $\varepsilon_{r\theta} < 0$.

2.1.2 Second case $\mu_{rz} < 0$

For $\varepsilon_{r\theta} > 0$, $\varepsilon_{r,eff}^{TE}$ is rewritten as

$$\varepsilon_{r,eff}^{TE} = |\varepsilon_{r\theta}| \left(1 + \frac{1}{|\varepsilon_{r\theta}\,\mu_{rz}| k_0^2} \cdot \left(\frac{u'_{nm}}{R}\right)^2\right) > 0. \tag{32}$$

And for $\varepsilon_{r\theta} < 0$, we obtain.

$$\begin{aligned} \varepsilon_{r,eff}^{TE} &= -|\varepsilon_{r\theta}| \left(1 - \frac{1}{|\varepsilon_{r\theta}\,\mu_{rz}| k_0^2} \cdot \left(\frac{u'_{nm}}{R}\right)^2\right) \\ &= -|\varepsilon_{r\theta}| \left(1 - \left(\frac{f_{c.nm}^{TE}}{f}\right)^2\right) > 0, \text{ if } f < f_{c.nm}^{TE}. \end{aligned} \tag{33}$$

Consequently, $\mu_{rz} < 0$ leads to $\varepsilon_{r,eff}^{TE} > 0$ below the cutoff frequency whenever $\varepsilon_{r\theta} > 0$ or $\varepsilon_{r\theta} < 0$.

Therefore, the relative permeability μ_{rz} below the cutoff frequency determines the sign of the relative effective permittivity of the anisotropic metamaterial in the circular waveguide. And the sign of the product $\mu_{rr}.\mu_{rz}$ of the metamaterial below the cutoff frequency determines the sign of the propagation constants of the waveguide studied.

The backward waves are obtained for $\mu_{rr} < 0$ and $\mu_{rz} > 0$ and the forward waves for $\mu_{rr} > 0$ and $\mu_{rz} < 0$ and. Therefore, the backward waves and the forward waves can propagate below the cutoff frequency.

2.2 Transverse magnetic (TM) modes

Similar to TE modes, TM modes can be derived as follows:
From Eq. (2), the differential equation for z-component can be obtained

$$\frac{\partial^2 E_z}{\partial r^2} + \frac{1}{r}\frac{\partial E_z}{\partial r} + \left(\frac{K_{c.r}^{(e)} \cdot \sqrt{\varepsilon_{r\theta}}}{K_{c.\theta}^{(e)} \cdot \sqrt{\varepsilon_{rr}}}\right)^2 \frac{1}{r^2}\frac{\partial^2 E_z}{\partial \theta^2} + \left(\frac{\sqrt{\varepsilon_{rz}}}{\sqrt{\varepsilon_{rr}}} K_{c.r}^{(e)}\right)^2 E_z = 0. \tag{34}$$

Using the separation of the variables (r,θ), the expression of the longitudinal electric field E_z for the TM_{nm} modes in the circular metallic waveguide completely filled with anisotropic metamaterial is necessary for the resolution of the differential Eq. (34). E_z can be written as follows

$$E_z^{(e)} = E_0 \cos\left(\frac{K_{c.\theta}^{(e)} \cdot \sqrt{\varepsilon_{rr}}}{K_{c.r}^{(e)} \cdot \sqrt{\varepsilon_{r\theta}}} n.\theta\right) J_n\left(\frac{\sqrt{\varepsilon_{rz}}}{\sqrt{\varepsilon_{rr}}} K_{c.r}^{(e)}.r\right) e^{-jk_z z} \tag{35}$$

The expressions (5)–(8) become

$$E_r^{(e)} = \frac{-jk_z}{K_{c.r}} \frac{\sqrt{\varepsilon_{rz}}}{\sqrt{\varepsilon_{rr}}} E_0 . \cos\left(\frac{K_{c.\theta}^{(e)} \cdot \sqrt{\varepsilon_{rr}}}{K_{c.r}^{(e)} \cdot \sqrt{\varepsilon_{r\theta}}} n.\theta\right) J'_n\left(\frac{\sqrt{\varepsilon_{rz}}}{\sqrt{\varepsilon_{rr}}} K_{c.r}^{(e)}.r\right) e^{-jk_z z} \tag{36}$$

$$E_\theta^{(e)} = \frac{jk_z}{K_{c.\theta}.K_{c.r}} \frac{\sqrt{\varepsilon_{rr}}}{\sqrt{\varepsilon_{r\theta}}} \frac{n}{r} E_0 . \sin\left(\frac{K_{c.\theta}^{(e)} \cdot \sqrt{\varepsilon_{rr}}}{K_{c.r}^{(e)} \cdot \sqrt{\varepsilon_{r\theta}}} n.\theta\right) J_n\left(\frac{\sqrt{\varepsilon_{rz}}}{\sqrt{\varepsilon_{rr}}} K_{c.r}^{(e)}.r\right) e^{-jk_z z} \tag{37}$$

$$H_r^{(e)} = \frac{-j\omega\varepsilon_0}{K_{c.\theta}.K_{c.r}} \sqrt{\varepsilon_{r\theta}}\sqrt{\varepsilon_{rr}}.\frac{n}{r}E_0.\sin\left(\frac{K_{c.\theta}^{(e)}.\sqrt{\varepsilon_{rr}}}{K_{c.r}^{(e)}.\sqrt{\varepsilon_{r\theta}}}n.\theta\right)J_n\left(\frac{\sqrt{\varepsilon_{rz}}}{\sqrt{\varepsilon_{rr}}}K_{c.r}^{(e)}.r\right)e^{-jk_z z} \tag{38}$$

$$H_\theta^{(e)} = \frac{-j\omega\varepsilon_0}{K_{c.r}} \sqrt{\varepsilon_{rr}}\sqrt{\varepsilon_{rz}}.E_0.\cos\left(\frac{K_{c.\theta}^{(e)}.\sqrt{\varepsilon_{rr}}}{K_{c.r}^{(e)}.\sqrt{\varepsilon_{r\theta}}}n.\theta\right)J'_n\left(\frac{\sqrt{\varepsilon_{rz}}}{\sqrt{\varepsilon_{rr}}}K_{c.r}^{(e)}.r\right)e^{-jk_z z} \tag{39}$$

The boundary condition (18) gives the following equation

$$J_n(u_{nm}) = 0. \tag{40}$$

with

$$u_{nm} = \frac{\sqrt{\varepsilon_{rz}}}{\sqrt{\varepsilon_{rr}}}K_{c.r}^{(e)}.R. \tag{41}$$

In Eq. (41) u_{nm} represents the mth zero (m = 1, 2, 3, …) of the Bessel function J_n of the first kind of order n.

The constant E_0 is determined by normalizing the power flow down the circular guide.

$$P^{TM} = \int_0^R\int_0^{2\pi}\left(E_r^{(e)}H_\theta^{*(e)} - E_\theta^{(e)}H_r^{*(e)}\right)rdrd\theta = 1 \tag{42}$$

Eq. (42) gives:

$$E_0 = \frac{K_{c.r}^2}{\sqrt{\omega\varepsilon_0\varepsilon_{rr}k_z}}N_{nm}^{(e)} \tag{43}$$

with

$$N_{nm}^{(e)} = \frac{1}{u_{nm}.J'_n(u_{nm}).\sqrt{\frac{\delta_n}{2}}} \tag{44}$$

$$\delta_n = \begin{cases} 2\pi, \ if \ n = 0 \\ \pi - \dfrac{\sin(4\pi b.\mathrm{n})}{4\mathrm{b.n}}, \ if \ n > 0 \end{cases} \tag{45}$$

$$b = \frac{K_{c.\theta}^{(e)}.\sqrt{\varepsilon_{rr}}}{K_{c.r}^{(e)}.\sqrt{\varepsilon_{r\theta}}} \tag{46}$$

Finally, the propagation constant in TM mode is given by:

$$k_{z.nm}^{(TM)} = \pm\sqrt{k_0^2\varepsilon_{rr}.\mu_{r\theta} - \frac{\varepsilon_{rr}}{\varepsilon_{rz}}\left(\frac{u_{nm}}{R}\right)^2} \tag{47}$$

Obviously, the cutoff frequency is written

$$f_{c.nm}^{(TM)} = \frac{c}{2\pi}\frac{1}{\sqrt{|\mu_{r\theta}\varepsilon_{rz}|}}.\left(\frac{u_{nm}}{R}\right). \tag{48}$$

We can introduce the following effective permeability and effective permittivity to describe the propagation characteristics of the waveguide modes.

$$\varepsilon_{r,eff}^{TM} = \varepsilon_{rr}, \tag{49}$$

$$\mu_{r,eff}^{TM} = \mu_{r\theta}\left(1 - \frac{1}{\mu_{r\theta}\varepsilon_{rz}k_0^2}\cdot\left(\frac{u_{nm}}{R}\right)^2\right). \tag{50}$$

Similar to the previous discussion, we have three possibilities:
Further, It is apparent that:

- $k_z^{TM} = k_0\sqrt{\mu_{r,eff}^{TM}.\varepsilon_{r,eff}^{TM}} > 0$, for $\mu_{r,eff}^{TM} > 0$ and $\varepsilon_{r,eff}^{TM} > 0$;
- $k_z^{TM} = -k_0\sqrt{\mu_{r,eff}^{TM}.\varepsilon_{r,eff}^{TM}} < 0$, for $\mu_{r,eff}^{TM} < 0$ and $\varepsilon_{r,eff}^{TM} < 0$;
- $k_z^{TM} = \pm jk_0\sqrt{\mu_{r,eff}^{TM}.\varepsilon_{r,eff}^{TM}}$, for $\mu_{r,eff}^{TM}.\varepsilon_{r,eff}^{TM} < 0$.

Consequently, the sign of $\mu_{r,eff}^{TM}$ depends on the sign of ε_{rz}. In the following, we will consider all cases that arise from the different sign of ε_{rz}.

2.2.1 Case when $\varepsilon_{rz} > 0$

In this case, for $\mu_{r\theta} > 0$, $\mu_{r,eff}^{TM}$ is rewritten as.

$$\mu_{r,eff}^{TM} = |\,\mu_{r\theta}|\left(1 - \frac{1}{|\,\mu_{r\theta}\ \varepsilon_{rz}|k_0^2}\cdot\left(\frac{u_{nm}}{R}\right)^2\right) = |\,\mu_{r\theta}|\left(1 - \left(\frac{f_{c.nm}^{TM}}{f}\right)^2\right) < 0,\ \text{if}\, f < f_{c.nm}^{TM} \tag{51}$$

And for $\mu_{r\theta} < 0$, we have

$$\mu_{r,eff}^{TM} = -|\,\mu_{r\theta}|\left(1 + \frac{1}{|\,\mu_{r\theta}\varepsilon_{rz}|k_0^2}\cdot\left(\frac{u_{nm}}{R}\right)^2\right) < 0. \tag{52}$$

It can be seen that $\varepsilon_{rz} > 0$ leads to $\mu_{r,eff}^{TM} < 0$ below the cutoff frequency whenever $\mu_{r\theta} > 0$, or $\mu_{r\theta} < 0$.

2.2.2 Case when $\varepsilon_{rz} < 0$

In this case, for $\mu_{r\theta} > 0$, we have

$$\mu_{r,eff}^{TM} = |\,\mu_{r\theta}|\left(1 + \frac{1}{|\,\mu_{r\theta}\ \varepsilon_{rz}|k_0^2}\cdot\left(\frac{u_{nm}}{R}\right)^2\right) > 0. \tag{53}$$

and for $\mu_{r\theta} < 0$, we obtain.

$$\mu_{r,eff}^{TM} = -|\,\mu_{r\theta}|\left(1 - \frac{1}{|\,\mu_{r\theta}\ \varepsilon_{rz}|k_0^2}\cdot\left(\frac{u_{nm}}{R}\right)^2\right) = -|\,\mu_{r\theta}|\left(1 - \left(\frac{f_{c.nm}^{TM}}{f}\right)^2\right) > 0.,\ \text{if}\, f < f_{c.nm}^{TM} \tag{54}$$

It is also seen that the relative permittivity ε_{rz} which is independent of $\mu_{r\theta}$ determines the sign of the relative effective permeability $\mu_{r,eff}^{TM}$ of the anisotropic metamaterial in the circular waveguide. The forward wave propagates in the waveguide for $\varepsilon_{rz} < 0$ and $\varepsilon_{rr} > 0$, and backward wave propagates for $\varepsilon_{rz} > 0$ and $\varepsilon_{rr} < 0$.

Therefore from this analysis, it is found that both the backward waves and the forward waves can propagate in any frequency region. This is determined by the sign of ε_{rz} and ε_{rr} for TM modes and the sign of μ_{rz} and μ_{rr} for TE modes.

2.3 Analysis of uniaxial discontinuities in the circular waveguides

In this section, we analyzed a waveguide filters filled with partially anisotropic metamaterial using the extension of the mode matching technique based on the Scattering Matrix Approach which, from the decomposition of the modal fields, are used to determine the dispersion matrix and thus the characterization of a discontinuity in waveguide. The discontinuities are considered without losses.

In **Figure 2** we consider a junction between two circular waveguides having the same cross section filled with two different media. a^i and $b^i$are the incident and the reflected waves, respectively.

The transverse electric and magnetic fields (E_T, H_T) in the wave guides can be written in the modal bases as follows [20]:

$$E_T = \sum_{m=1}^{\infty} A_m^i (a_m^i + b_m^i) e_m^i \tag{55}$$

$$H_T = \sum_{m=1}^{\infty} B_m^i (a_m^i - b_m^i) h_m^i \tag{56}$$

where H_T and E_T are the transverse magnetic and electric fields (T refers to the components in the transverse plane), h_m^i, e_m^i represent the mth magnetic and electric modal Eigen function in the guide i, respectively and A_m^i and B_m^i are complex coefficients which are determined by normalizing the power flow down the circular guides (m is the index of the mode and $i = I, II$).

At the junction, the continuity of the fields allows to write the following equations:

$$E_t^I = E_t^{II} \tag{57}$$

$$H_t^I = H_t^{II} \tag{58}$$

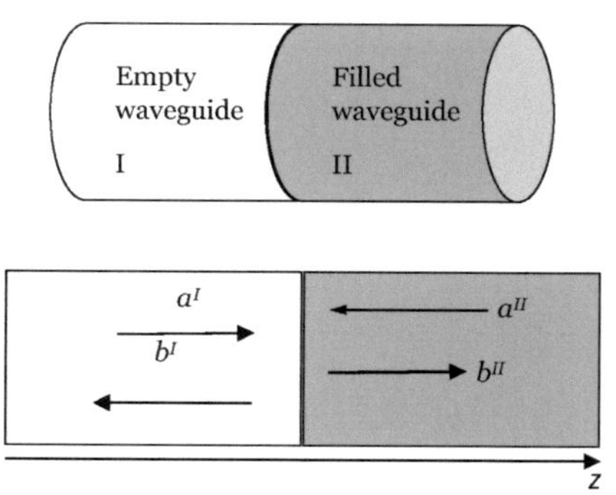

Figure 2.
Junction between two circular waveguides filled with two different media having the same cross section.

By postponing the Eqs. (55) and (56) in (57) and (58), we obtain:

$$\sum_{m=1}^{N_1} A_m^I (a_m^I + b_m^I) e_m^I = \sum_{p=1}^{N_2} A_p^{II} \left(a_p^{II} + b_p^{II} \right) e_p^{II} \tag{59}$$

$$\sum_{m=1}^{N_1} B_m^I (a_m^I - b_m^I) h_m^I = \sum_{p=1}^{N_2} B_p^{II} \left(-a_p^{II} + b_p^{II} \right) h_p^{II} \tag{60}$$

N_1 and N_2 are the number of considered modes in guides 1 and 2, respectively. By applying the Galerkin method, Eqs. (59) and (60), lead to the following systems:

$$\sum_{m=1}^{N_1} A_m^I (a_m^I + b_m^I) \left\langle e_m^I | e_p^{II} \right\rangle = A_p^{II} \left(a_p^{II} + b_p^{II} \right) \tag{61}$$

$$B_m^I (a_m^I - b_m^I) = \sum_{p=1}^{N_2} B_p^{II} \left(-a_p^{II} + b_p^{II} \right) \left\langle h_p^{II} | h_m^I \right\rangle \tag{62}$$

The inner product is defined as:

$$\langle e_m | e_p \rangle = \int_S e_m^* e_p \, dS \tag{63}$$

The Eqs. (61) and (62) give:

$$-a_p^{II} + \sum_{m=1}^{N_1} \frac{A_m^I}{A_p^{II}} a_m^I \left\langle e_m^I | e_p^{II} \right\rangle = b_p^{II} - \sum_{m=1}^{N_1} \frac{A_m^I}{A_p^{II}} b_m^I \left\langle e_m^I | e_p^{II} \right\rangle \tag{64}$$

$$a_m^I + \sum_{p=1}^{N_2} \frac{B_p^{II}}{B_m^I} a_p^{II} \left\langle h_p^{II} | h_m^I \right\rangle = b_m^I + \sum_{p=1}^{N_2} \frac{B_p^{II}}{B_m^I} b_p^{II} \left\langle h_p^{II} | h_m^I \right\rangle \tag{65}$$

which can be written in matrix form:

$$\begin{bmatrix} U & M_1 \\ M_2 & -U \end{bmatrix} \begin{bmatrix} a_1^I \\ . \\ a_{N_1}^I \\ a_1^{II} \\ . \\ a_{N_2}^{II} \end{bmatrix} = \begin{bmatrix} U & M_1 \\ -M_2 & U \end{bmatrix} \begin{bmatrix} b_1^I \\ . \\ b_{N_1}^I \\ b_1^{II} \\ . \\ b_{N_2}^{II} \end{bmatrix} \tag{66}$$

where U is the identity matrix. M_1 and M_2 are defined as:

$$M_{1ij} = \frac{B_j^{II}}{B_i^I} \left\langle h_j^{II} | h_i^I \right\rangle \tag{67}$$

$$M_{2ij} = \frac{A_i^I}{A_j^{II}} \left\langle e_i^I | e_j^{II} \right\rangle \tag{68}$$

The scattering matrix of the discontinuity is:

$$S = \begin{bmatrix} U & M_1 \\ -M_2 & U \end{bmatrix}^{-1} \begin{bmatrix} U & M_1 \\ M_2 & -U \end{bmatrix} \tag{69}$$

The total scattering matrix is obtained by chaining the S scattering matrices of all the discontinuities in a waveguide having cascaded uniaxial discontinuities [21].

3. Numerical results and discussion

3.1 Propagating modes

We choose the radius of the circular metal guide R = 13.4 mm.

In a first case, we study the TE modes of a circular guide completely filled with anisotropic metamaterials (see **Figure 1**) with negative μ_{rr} or negative μ_{rz}. The fundamental mode of the equivalent empty circular waveguide has a resonant frequency of 6.57 GHz. For the case of metamaterials with a permeability $\mu_r = -1$ and permittivity $\varepsilon_r = -4.4$, the fundamental mode presents a resonance frequency of $f_{c.11}^{TE} = 3.13\ GHz$.

In **Figure 3** the curves of the propagation constant, for frequency range 1–10 GHz and for the first five TE modes with $\mu_{rr} = 1$, $\mu_{rz} = -1$ and $\varepsilon_{r\theta} = 4.4$, are represented. We observe that all modes propagate without cutoff frequencies (forward waves). **Figure 4** represents the same diagrams for $\mu_{rr} = -1$, $\mu_{rz} = 1$ and $\varepsilon_{r\theta} = 4.4$. When n and m are small and ω is large, the waves stop propagating. So, these modes propagate at low frequencies and cutoff at high frequencies (backward waves).

It is interesting to see that both forward and backward waves can be obtained by controlling the signs of μ_{rz} and μ_{rr}. Our results agree well with the predicted ones.

In a second case, we study the TE modes of this circular waveguide. **Figure 5** represents the curves of propagation constant for the frequency range 1–10 GHz

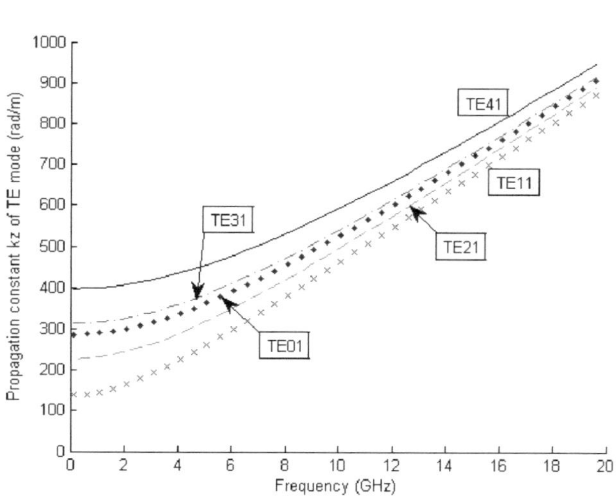

Figure 3.
Curves of propagation constant k_z^{TE} *for TE mode of the circular waveguide completely filled anisotropic metamaterial with parameters* $\mu_{rr} = 1$, $\mu_{rz} = -1$, $\varepsilon_{r\theta} = 4.4$.

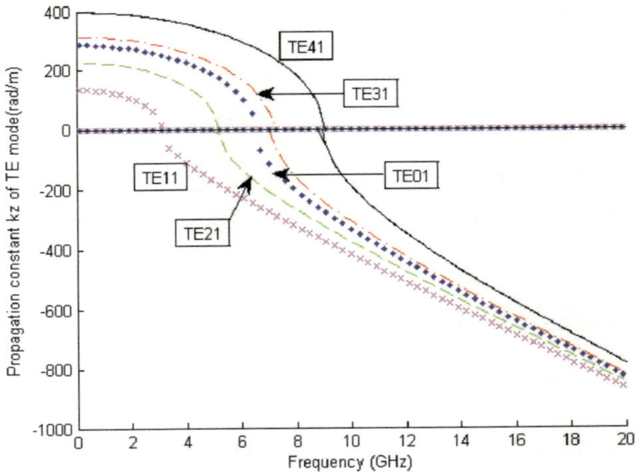

Figure 4.
Curves of propagation constant k_z^{TE} *for TE mode of the circular waveguide completely filled anisotropic metamaterial with parameters* $\mu_{rr} = -1$, $\mu_{rz} = 1$ *and* $\varepsilon_{r\theta} = 4.4$.

and for the first five TM modes with $\varepsilon_{rr} = 4.4$, $\varepsilon_{rz} = -4.4$ and $\mu_{r\theta} = 1$. All modes propagate without cutoff (forward waves).

Calculated curves of propagation constant for the frequency range 1–10 GHz and for the first five TM modes with $\varepsilon_{rr} = -4.4$, $\varepsilon_{rz} = 4.4$, $\mu_{r\theta} = 1$ are presented. We notice that both forward wave and backward wave can be obtained by controlling the signs of ε_{rr} and ε_{rz}. **Figures 5** and **6** show that our results agree well with the predicted ones.

We observe that the cutoff frequencies of lowest TE modes decreased with the respect increase of μ_{rz} for $\mu_{rr} = -1$ and $\varepsilon_{r\theta} = 4.4$ (see **Figure 7**). In a same manner, the TM cutoff frequencies decreased with the respect increase of ε_{rz} for $\varepsilon_{rr} = -4.4$ and $\mu_{r\theta} = 1$ (see **Figure 8**). Consequently, by varying the parameters of material the propagating mode can be controlled.

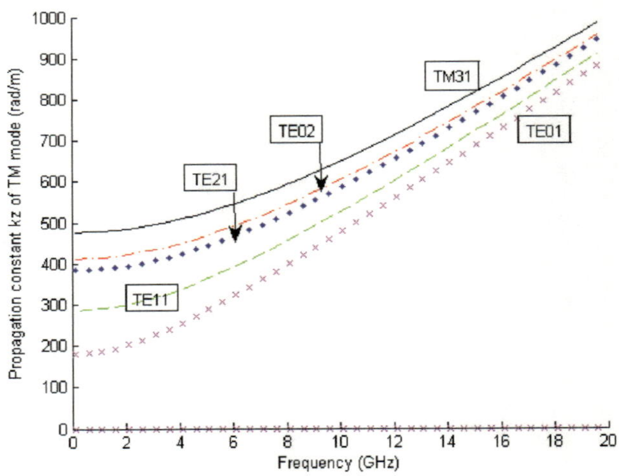

Figure 5.
Curves of propagation constant k_z^{TM} *for TM mode of the circular waveguide completely filled anisotropic metamaterial with parameters* $\varepsilon_{rr} = 4.4$, $\varepsilon_{rz} = -4.4$, $\mu_{r\theta} = 1$.

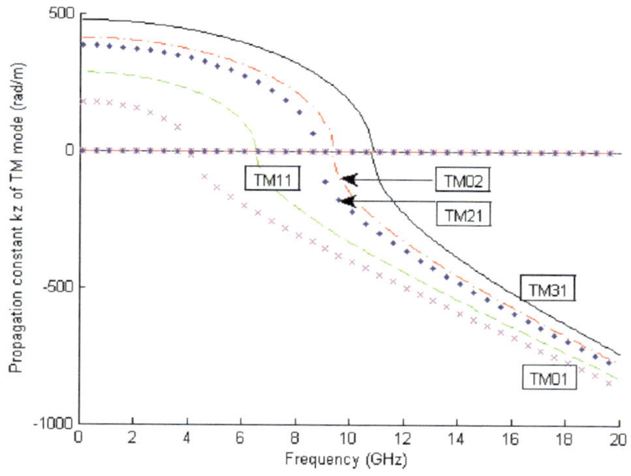

Figure 6.
Curves of propagation constant k_z^{TM} *for TM mode of the circular waveguide completely filled anisotropic metamaterial with parameters* $\varepsilon_{rr} = -4.4$, $\varepsilon_{rz} = 4.4$, $\mu_{r\theta} = 1$.

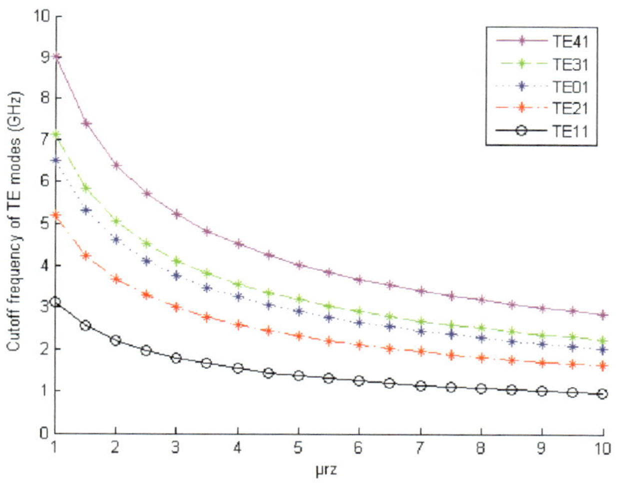

Figure 7.
The cutoff frequencies for the first five TE modes versus μ_{rz} *with* $\mu_{rr} = -1$, $\varepsilon_{r\theta} = 4.4$.

3.2 Filter design

We consider now, 12 discontinuities (see **Figure 9**) constituted by juxtaposing 13 circular waveguides having the same dimensions (R = 13.4 mm). The circuit is formed by alternation of empty guide ($\varepsilon_r = \mu_r = 1$) of width l = 10 mm and guide filled by anisotropic metamaterials ($\varepsilon_{rr} = \varepsilon_{r\theta} = -\varepsilon_{rz} = -4.4$; $\mu_{rr} = \mu_{r\theta} = -\mu_{rz} = 1$) of width d = 0.2 mm (periodic structure). **Figure 9** represents the geometry of the studied structure.

The transmission and reflection coefficients using our numerical method with MATLAB and HFSS are presented in **Figure 10**. We used 8 modes in the whole circuit for the modal method. The simulations results show that are in perfect agreement. However and especially if the number of discontinuities increases, our method is significantly faster than HFSS. Then, by using our approach, it could easy to design filters according to a given specifications.

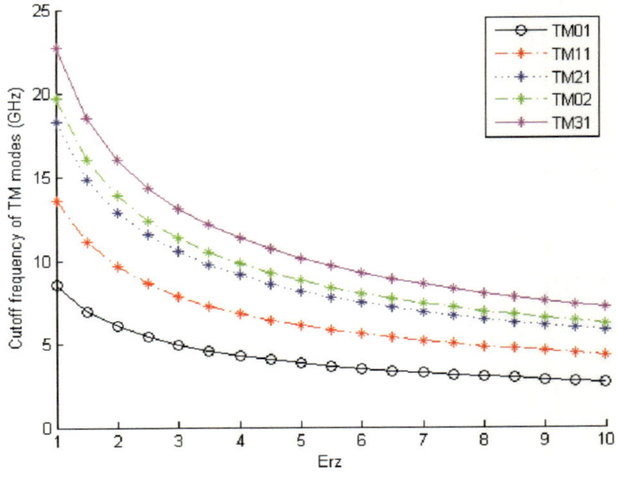

Figure 8.
The cutoff frequencies for the first five TM modes versus ε_{rz} *with* $\varepsilon_{rr} = -4.4$, $\mu_{r\theta} = 1$.

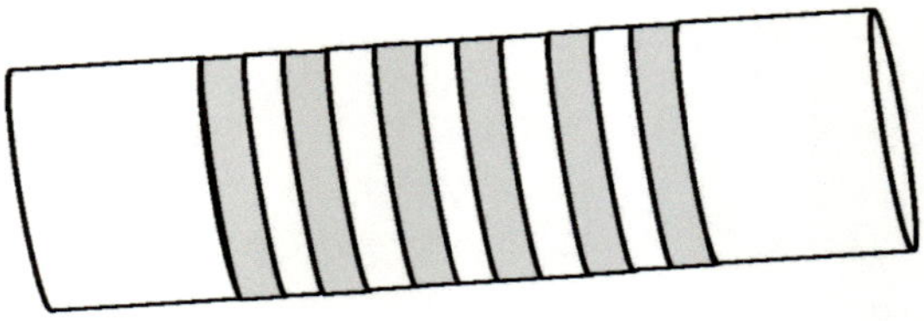

Figure 9.
Geometry of the circular waveguide with 12 discontinuities.

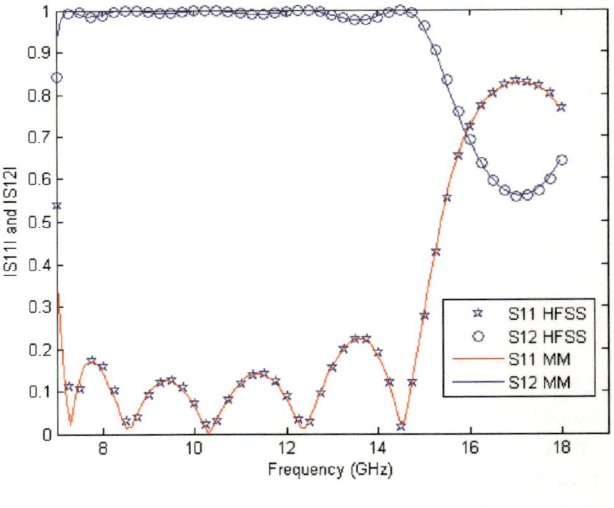

Figure 10.
Reflection coefficient of the periodic structure with 12 discontinuities.

4. Conclusion

Rigorous analysis of propagating modes in circular waveguides filled with anisotropic metamaterial has been developed. It was demonstrated that the propagation constant of the waveguide are closely dependent on constitutive parameters of the metamaterial. Using our MATLAB code the dispersion curves of the fundamental mode and the first four higher order modes of the metamaterial waveguide are obtained.

We found that in different frequency ranges below and above the cutoff frequency both the forward and the backward waves can propagate. This is determined by the sign of ε_{rz} and ε_{rr} for TM modes and by the sign of μ_{rz} and μ_{rr} for TE modes. Our simulation results are in good agreement with the theoretical prediction.

Moreover, using the Scattering Matrix Approach we applied the extension of MM technique to determine the dispersion matrix and to analyze multiple uniaxial circular discontinuity in waveguide filled with anisotropic metamaterials. This introduced tool is applied to the modeling of large complex structures such as filters where its rapidity compared to the commercial simulation tools is verified.

Author details

Hedi Sakli[1,2]* and Wyssem Fathallah[1]

1 MACS Research Laboratory, National Engineering School of Gabes, Gabes University, Gabes, Tunisia

2 EITA Consulting, Montesson France

*Address all correspondence to: hedi.s@eitaconsulting.fr

References

[1] Alu A, Engheta N. Mode excitation by a line source in a parallel plate waveguide filled with a pair of parallel double-negative and double-positive slabs. In: IEEE AP-S International Symposium. Columbus, OH; 2003. pp. 359-362

[2] Xu Y. Wave propagation in rectangular waveguide filled with single negative metamaterial slab. Electronics Letters. 2003;**39**(25):1831-1833

[3] Xu Y. A study of waveguides field with anisotropic metamaterials. Microwave and Optical Technology Letters. 2004;**41**:426-431

[4] Alu A, Engheta N. Guided modes in a waveguide filled with a pair of single-negative (SNG), double negative (DNG), and/or double-positive (DPS) layers. IEEE Transactions on Microwave Theory and Techniques. 2004;**52**: 199-210

[5] Cory H, Shtrom A. Wave propagation along a rectangular metallic waveguide longitudinally loaded with a metamaterial slab. Microwave and Optical Technology Letters. 2004;**41**(2): 123-127

[6] Meng FY, Wu Q, Fu JH, Gu XM, Li LW. Transmission characteristics of wave modes in a rectangular waveguide filled with anisotropic metamaterial. Applied Physics A: Materials Science & Processing. 2009;**94**:747-753

[7] Meng FY, Wu Q, Li LW. Controllable Metamaterial-loaded waveguides supporting backward and forward waves transmission characteristics of wave modes in a rectangular waveguide filled with anisotropic metamaterial. IEEE Transactions on Antennas and Propagation. 2011;**59**(9):3400-3411

[8] Zhang D, Ma J. The propagation and cutoff frequencies of the rectangular metallic waveguide partially filled with metamaterial multilayer slabs. Progress In Electromagnetics Research (PIER) M. 2009;**9**:35-40

[9] Pan Y, Xu S. Cpmlex modes in parallel-plate waveguide structure filled with left-handed material. Chinese Journal of Electronics. 2009;**18**(3): 551-554

[10] Cojocaru E. Waveguides filled with bilayers of double-negative (DNG) and double-positive (DPS) metamaterials. Progress In Electromagnetics Research (PIER) B. 2011;**32**:75-90

[11] Fathallah W, Sakli H, Aguili T. Electromagnetic wave propagation in anisotropic Metamaterial waveguides. International Journal of Numerical Modelling: Electronic Networks, Devices and Fields. 2015;**28**(4):479-486

[12] Fathallah W, Sakli H, Aguili T. Full-wave study of rectangular metamaterial waveguides using Galerkin's method. In: 8th International Multi-Conference on Systems, Signals and Devices, SSD'13, March 2013. Hammamet, Tunisia;

[13] Marques R, Martel J, Mesa F, Medina F. Left-handed-media simulation and transmission of EM waves in subwavelength split-ring-resonator-loaded metallic waveguides. Physical Review Letters. 2002;**89**:183901

[14] Balanis CA. Circular waveguides. In: Material in Advanced Engineering Electromagnetics. Vol. 9. New York: Wiley; 1989. pp. 643-650

[15] Boyenga DL, Mabika CN, Diezaba A. A new multimodal Variational formulation analysis of cylindrical waveguide uniaxial discontinuities. Research Journal of Applied Sciences, Engineering and Technology. 2013;**6**(5):787-792

[16] Thabet R, Riabi ML, Belmeguenai M. Rigorous design and efficient optimization of quarter-wave transformers in metallic circular waveguides using the mode-matching method and the genetic algorithm. Progress in Electromagnetics Research. 2007;**68**:15-33

[17] Mahmoud SF. Guided modes on open chirowaveguides. IEEE Transactions on Microwave Theory and Techniques. 1995;**43**(1):205-209

[18] Shadrivov IV, Sukhorukov AA, Kivshar YS. Guided modes in negative-refractive-index waveguides. Physical Review E. 2003;**67**:057602

[19] Dong J-F, Li J. Characteristics of guided modes in uniaxial chiral circular waveguides. Progress in Electromagnetics Research. 2012;**124**: 331-345

[20] Couffignal P. Contribution à l'étude Des Filters en Guides métalliques [thesis]. INP Toulouse; 1992

[21] Ghosh B, Kakade AB. Guided modes in a metamaterial-filled circular waveguide. Electromagnetics. 2012;**32**: 465-480

Chapter 6

Long-Distance and Low-Radiation Waveguide Antennas for Wireless Communication Systems inside Tunnels

Shotaro Ishino, Satoshi Denno, Narumi Yashiro and Seiichi Suzuki

Abstract

Wireless LAN usage is also increasing at construction and civil engineering sites, and the efficiency of ICT construction has increased due to the use of tablet PCs and network cameras. When constructing a wireless LAN environment, for example, a LAN cable may be laid from outside the tunnel, and a number of wireless access points (APs) may be installed. However, it is not advantageous to use a large number of APs because the system price increases significantly. We consider using a long leaky-wave antenna to provide one AP. The reason for using a leaky-wave antenna is that, since the total tunnel length is on the order of km, it is necessary to reduce the power radiated by the antenna as much as possible to provide a functional communication area over a long distance. To reduce such transmission losses, we used a waveguide. A waveguide is a low-loss line and can function as a low-loss and low-radiation leaky-wave antenna which is suitable for long-distance communications; this is accomplished by combining a waveguide with a low-radiation antenna mechanism. In this chapter, we report the development of a waveguide-type leaky-wave antenna and the development of a wireless LAN environment in a tunnel.

Keywords: wireless communication, microwave, waveguide antenna, long distance, low radiation

1. Introduction

In recent years, wireless LANs have become widespread and are indispensable for convenient Internet use. Wireless LAN usage is also increasing at construction and civil engineering sites, and the efficiency of ICT construction has increased due to the use of tablet PCs and network cameras. Remote monitoring of worksite interiors contributes greatly to more rapid construction. This means that construction of wireless environments in the field is indispensable. Tunnel construction often takes place in the mountains. An LTE (4G) line may become disconnected but it cannot connect at all to the line in the tunnel.

This is because it is difficult to transmit radio waves from outside the tunnel to the inside. Radio waves are reflected and absorbed by thick soil and concrete walls, and this means that no external radio waves can be received inside the tunnel [1]. Therefore, it is common to lay a wired line from the outside of the tunnel and to construct a telephone line or other types of communication line in the tunnel. When constructing a wireless LAN environment, for example, a LAN cable may be laid from outside the tunnel and a number of wireless access points (APs) may be installed. However, working with a wired line is complicated, and there is a high risk of disconnection due to contact with building materials and construction equipment. Therefore, a high degree of robustness with fewer system failures is also required. Also, it is not advantageous to use a large number of APs because the system price increases significantly.

Therefore, we consider using a long leaky-wave antenna to provide one AP. The reason for using a leaky-wave antenna is that, since the total tunnel length is on the order of km, it is necessary to reduce the power radiated by the antenna as much as possible to provide a functional communication area over a long distance. The use of leaky coaxial cable (LCX) as a leaky-wave antenna has been studied [2–6]. As shown in **Figure 1**, a long-range communication area can be constructed by reradiating the output of the AP from a long LCX. However, in practice, the available power is attenuated in the cable due to transmission losses (mainly dielectric losses) of the LCX, so it is difficult to communicate over long distances.

To reduce such transmission losses, we used a waveguide [7]. A waveguide is a low-loss line and can function as a low-loss and low-radiation leaky-wave antenna which is suitable for long-distance communications; this is accomplished by combining a waveguide with a low-radiation antenna mechanism. In this paper, we report the development of a waveguide-type leaky-wave antenna and the development of a wireless LAN environment in a tunnel. We evaluated the system in the W56 band (5.5–5.7 GHz), as specified by IEEE 802.11 "a", "n" and shown in **Figure 2**.

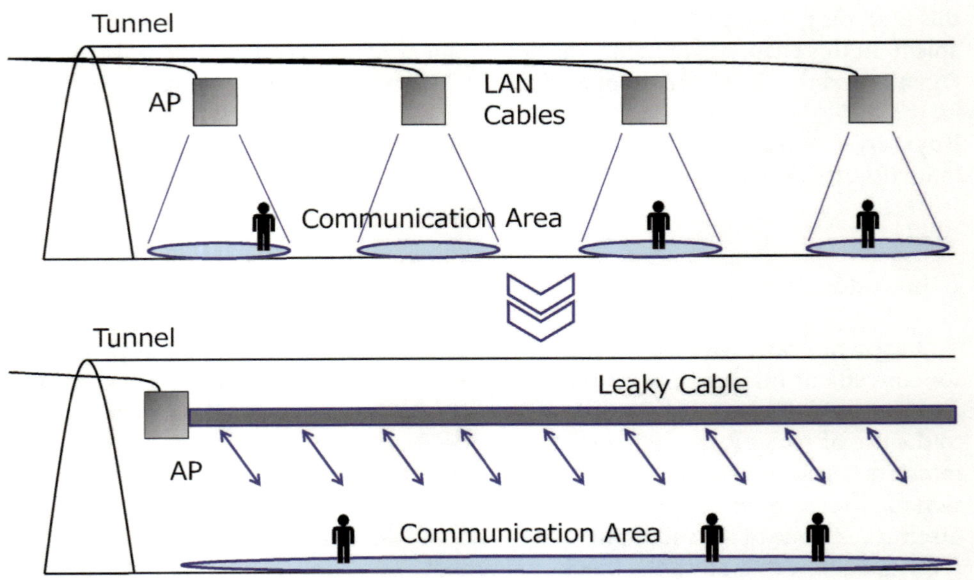

Figure 1.
Wireless communication system using a leaky cable.

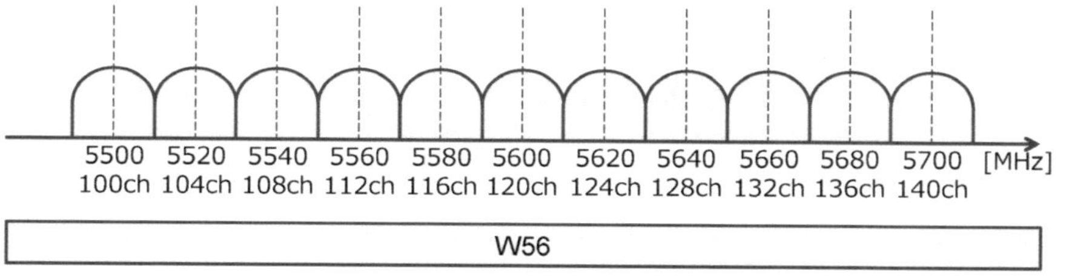

Figure 2.
WLAN W56 band.

2. Overview of the waveguide antenna

A waveguide is a hollow metal tube and transmits radio waves using reflections within the tube. In particular, rectangular waveguides are used in various applications such as radar, microwave oven, and microwave feeds. Transmission losses are low even in a microwave band or in a millimeter-wave band, and waveguides are expected to be used as transmission lines for next-generation communications.

A waveguide has a rectangular cross section as shown in **Figure 3**, and radio waves are transmitted by reflecting between the two metal plates on both ends of the waveguide at an angle θ, as shown in **Figure 4**. **Figure 3** shows a waveguide

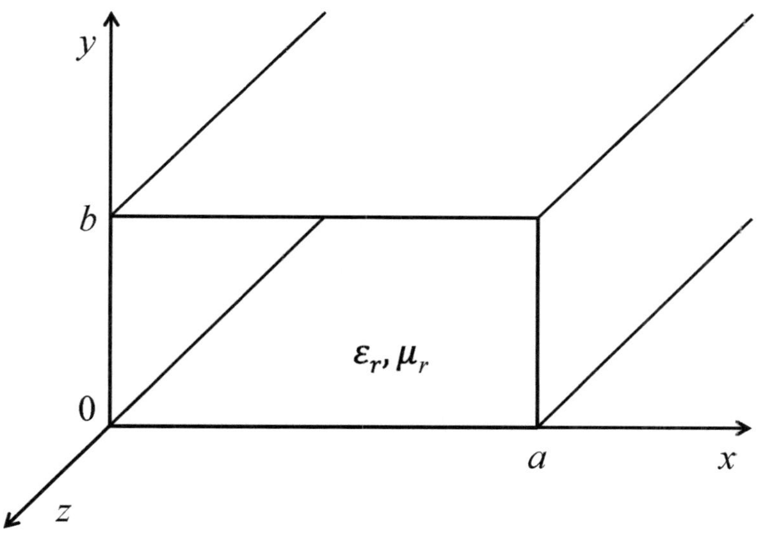

Figure 3.
Rectangular waveguide.

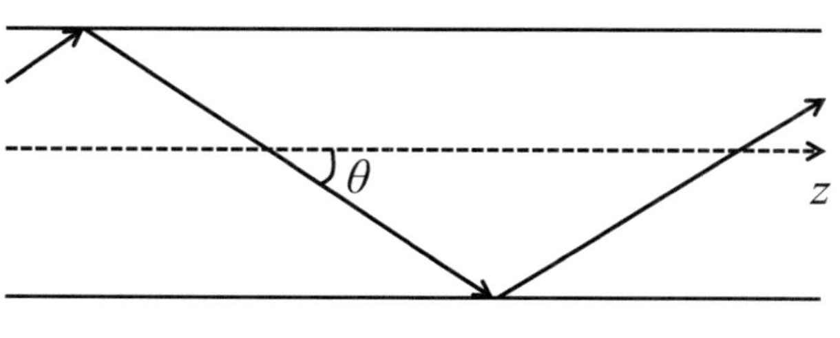

Figure 4.
Microwaves transmitted while reflecting between two metal plates.

filled with a material and covered entirely with a conductor. The relative permittivity of the material is ε_r, the relative permeability is μ_r, and the cross-sectional dimension of the waveguide is a × b.

Since the reflected waves at both ends are combined into one wave in the waveguide, the wave is transmitted within the waveguide at the guide wavelength λ_g. Also, as the frequency decreases, the angle θ increases, and the radio wave stops traveling back and forth between the metal plates on both ends when θ = 90°. The frequency at which this condition occurs is called the cutoff frequency f_c, and the free-space wavelength of f_c is called the cutoff wavelength λ_c. Eqs. 1–3 show these relationships.

$$\lambda_c = 2a\sqrt{\varepsilon_r \mu_r} \quad (1)$$

$$f_c = \frac{3 \times 10^8}{\lambda_c} \quad (2)$$

$$\lambda_g = \frac{\lambda}{\cos\theta} = \frac{\lambda}{\sqrt{1 - \left(\frac{\lambda_0}{\lambda_c}\right)^2}} = \frac{\lambda}{\sqrt{1 - \left(\frac{f_c}{f}\right)^2}} \quad (3)$$

This paper deals with the TE_{10} mode, which is the most basic transmission mode. In this mode, the current and electric field shown in **Figure 5** flow through the waveguide. For this case, if a slot is provided that is orthogonal to the current direction, radiation is emitted from the slot, and a waveguide antenna is created.

A traveling-wave antenna is one that continuously emits energy by using a traveling wave; surface-wave antennas using a dielectric line and leaky-wave antennas using a waveguide or coaxial line are well known. As described above, a leaky coaxial cable (LCX) is utilized as a highly flexible traveling-wave antenna by creating a slot in the outer conductor. However, LCX is not suitable for long-distance applications due to its large dielectric losses in the 5 GHz band.

In the future, as the frequencies used shift to quasi-millimeter-wave or millimeter-wave communications, dielectric losses will increase. A waveguide is

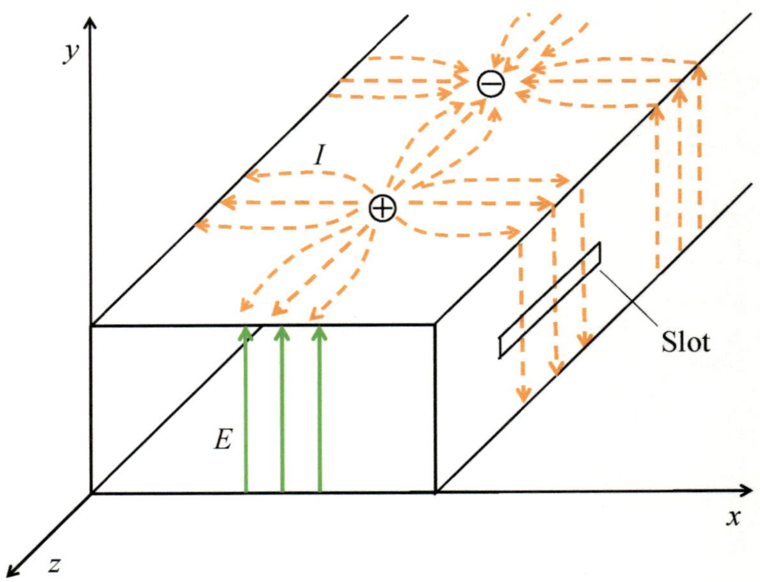

Figure 5.
Current and electric field flowing in a TE_{10} mode waveguide and slot configuration.

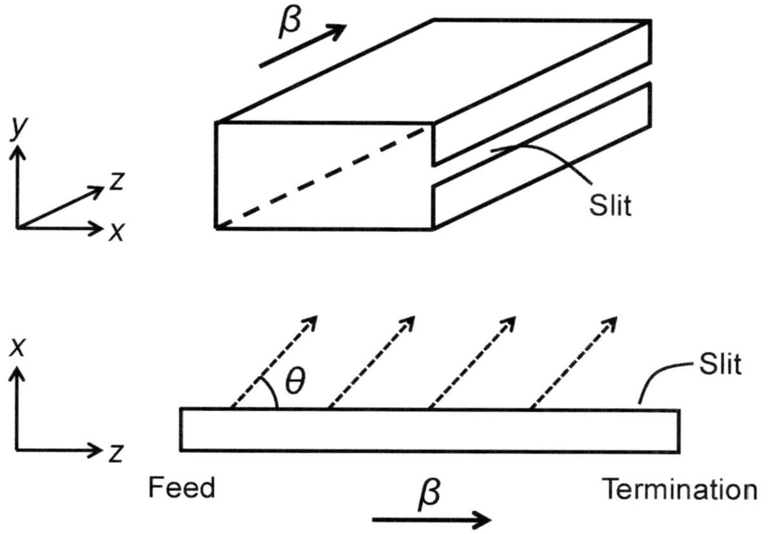

Figure 6.
Structure of a traveling-wave leaky waveguide.

considered to be useful because it has no dielectric losses and has small overall transmission losses. **Figure 6** shows a general traveling-wave leaky waveguide (with a phase constant $\beta > 0$) [8, 9]. In the leaky waveguide, the main beam is formed and emitted in the direction θ_{rad} for which the phases are aligned (Eq. 4). Here, v_0 is the free-space propagation velocity and v_g is the pipe propagation velocity.

When a continuous slit is present, the amount of radiation power increases, and the remaining power in the waveguide decreases. To apply this method to the proposed tunnel system, it is necessary to not only reduce the transmission losses but to also reduce the radiation levels and to maintain power in the pipe over long distances.

$$\theta_{rad} \cong \tan^{-1}\left(\frac{v_g}{v_0}\right) \tag{4}$$

3. Dual-plate leaky waveguide

The traveling-wave type of leaky waveguide shown in **Figure 6** requires a slit that is continuous in the direction of tube length, and such fabrication may increase the cost. To obtain a waveguide that is significantly cheaper than LCX, we need to minimize the number of processing steps for the tube. Therefore, we consider creating a waveguide (**Figure 7**) with a slit mechanism that is constructed by combining two U-shaped plates (metal plates). Bending such metal plates does not require a mold and they are easy to manufacture [10].

In the present waveguide, a slight gap is created between the plates, and the thickness is adjusted by sandwiching a thin insulating sheet or similar material. When viewed in the cross-sectional view, the present waveguide has the structure shown in **Figure 8(a)**, which is equivalent to the structure shown in **Figure 8(b)**. As shown in **Figure 9**, the results from the analysis of the transmission characteristics, as shown in **Figure 8(a)** and **(b)**, are nearly the same.

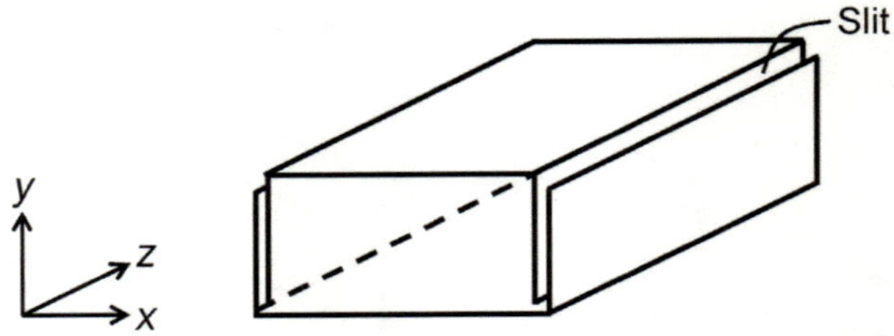

Figure 7.
Dual-plate leaky waveguide structure.

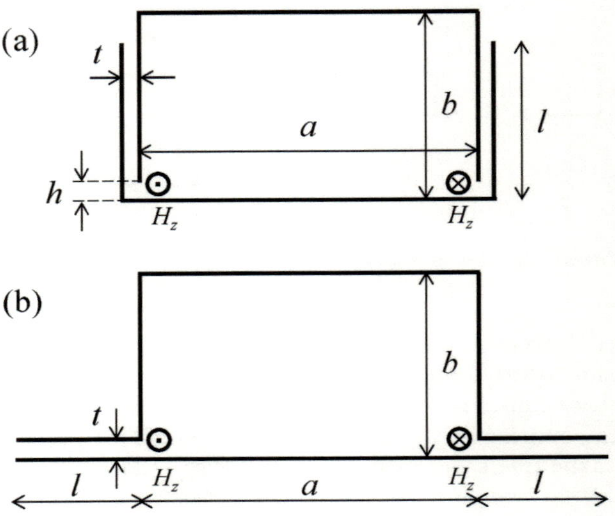

Figure 8.
(a) Cross section of leaky waveguide and (b) the equivalent cross section.

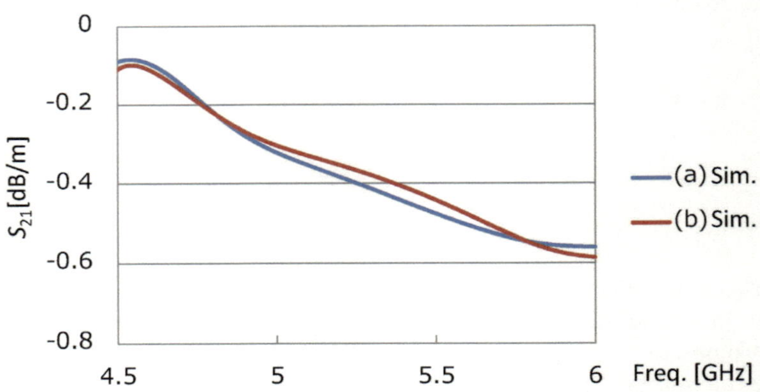

Figure 9.
(a, b) Analysis results of transmission characteristics.

Radiation is generated by the magnetic field which is generated in the opening of the waveguide. This is because the current flowing in the slit is determined by the H_z at that point where the slit entrance is short-circuited. However, since the direction of the magnetic field is reversed, the radiation level can be suppressed to a small value. When the slit length l is $\lambda/4$, the opening is short-circuited. Since the waveguide is a short-circuit boundary line, radiation theoretically does not occur.

Actually, since conductance is present in the opening, a short circuit does not occur but the radiation level is small.

When the impedance at the slit as viewed from inside the waveguide is low and when the slit height *h* in **Figure 8(a)** is sufficiently small, the TE_{10} mode is maintained. Since the slit satisfies the conditions of Eq. 5, only the TEM mode occurs. As *t* increases, the amount of radiation (and radiation resistance) increases because the magnetic current flowing to the surface when the slit entrance is short-circuited increases. **Table 1** shows the relevant design parameters.

The far-field radiation wave pattern and gain at 5.7 GHz of the designed waveguide were analyzed by simulation (Ansys HFSS). **Figures 10–12** show the results.

a	*b*	*t*	*h*	*l*	
40	20	0.05	0.05	20	[mm]

Table 1.
Design parameters.

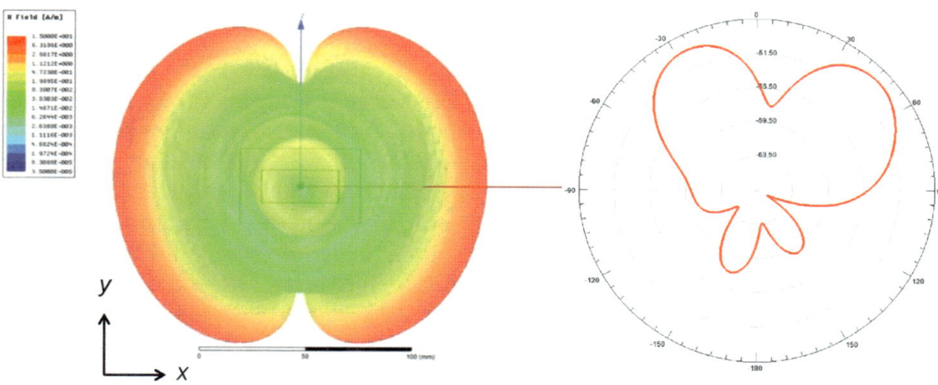

Figure 10.
Far-field radiation pattern and gain (x-y plane).

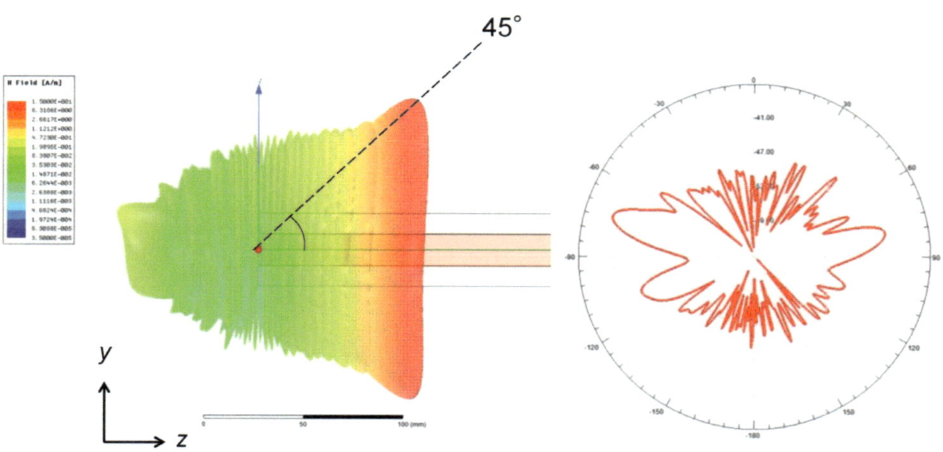

Figure 11.
Far-field radiation pattern and gain (y-z plane).

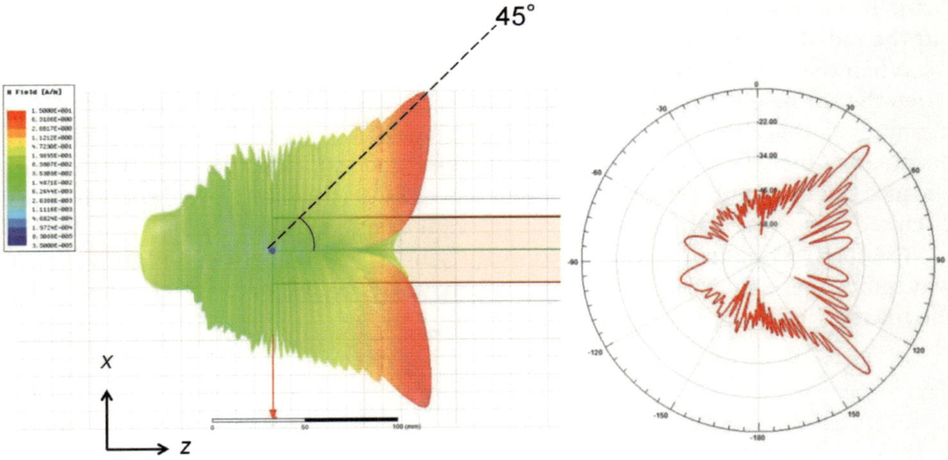

Figure 12.
Far-field radiation pattern and gain (z-x plane).

$$\frac{k}{k_c} = \frac{kt}{\pi} \ll 1 \tag{5}$$

The radiation directly above (on the *y*-axis) is canceled due to the phase inversion of the waves leaking from the left and right slits. It was also confirmed that the radiation was stronger at an angle of 45° with respect to the direction of propagation (e.g., *z*-axis direction). The maximum gain was −13.8 dBi. This is the property shown in **Figure 6**, and the radiation angle almost coincided with the calculated value. **Figure 13** shows the radiation state in the direction of propagation.

The waveguide must retain sufficient residual power over long distances. First, the conduction losses were examined. The conduction losses depend on the conductivity of the metal used for the waveguide, as shown in **Figure 14**. The waveguide discussed herein is made of aluminum, which has relatively good electrical conductivity; aluminum was used to manufacture the waveguide at a low cost. In addition, it is necessary to suppress the radiated power and reduce leakage. Therefore, whether the leakage can be reduced by adjusting the design parameters was examined. Here, it is assumed that there were no conductor losses.

First, the radiation loss when the slit height h is adjusted in the range of 0.05–0.5 mm is shown in **Figure 15** when t is set to 0.2 mm. In addition, to obtain the desired characteristics when the profile is decreased, the waveguide height was set to 10 mm without changing the waveguide width, and this case was evaluated. For this case, the slit length l was the same as b. It was found that h did not contribute to

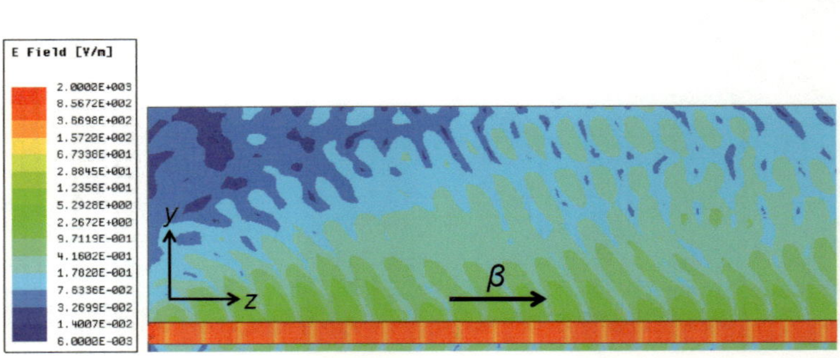

Figure 13.
Radiation in the direction of propagation (y-z plane).

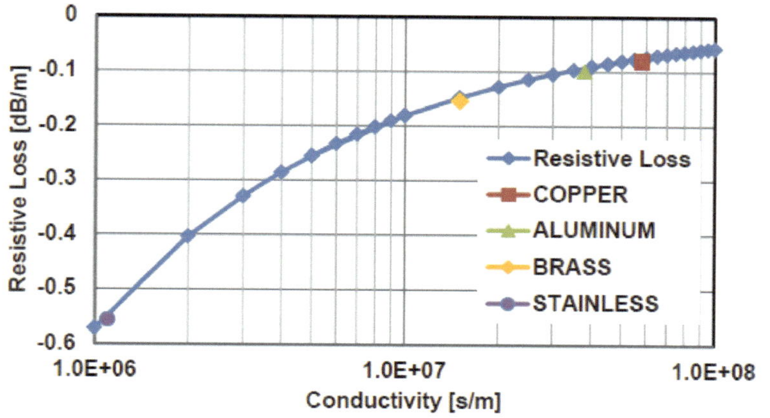

Figure 14.
Relationship between metal types and conduction losses.

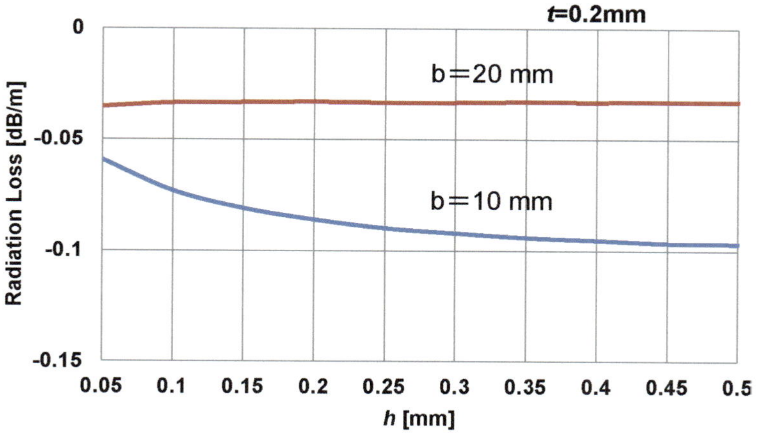

Figure 15.
Relationship between slit height and radiation.

radiation losses in this range. If h is smaller than the height of the waveguide, it is presumed that the TE_{10} mode does not change and that the H_z in the waveguide does not change. Therefore, the current flowing through the slit does not change.

In addition, the radiation increases due to the lower attitude. This is because the current increases. Although the degree of freedom and convenience of installation are improved by decreasing the waveguide profile, it has been confirmed that there is a trade-off. Naturally, when the profile is lower, the conductor losses for the waveguide increase. In this study, we focused on the radiation characteristics, but we also need to consider conductor losses.

Next, the radiation losses when the slit width t is similarly adjusted in the range of 0.05–0.5 mm are shown in **Figure 16**, and h is set to 0.2 mm. It was determined that the radiated power increased as t increased. For the case of the low attitude, the slit length l was also set to the same value as b, but the radiation also increased.

Since the characteristic impedance at the slit varies with Δt, it is necessary to consider conductor losses.

Finally, **Figure 17** shows the radiation losses when the slit length l is adjusted in the range of 0.5–20.0 mm for the conditions shown in **Table 1**. As l increases, the radiated power decreases. However, the change is small in the region of

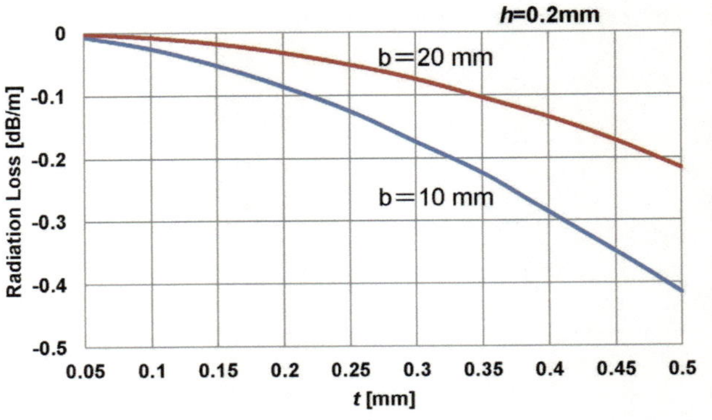

Figure 16.
Relationship between slit width and radiation.

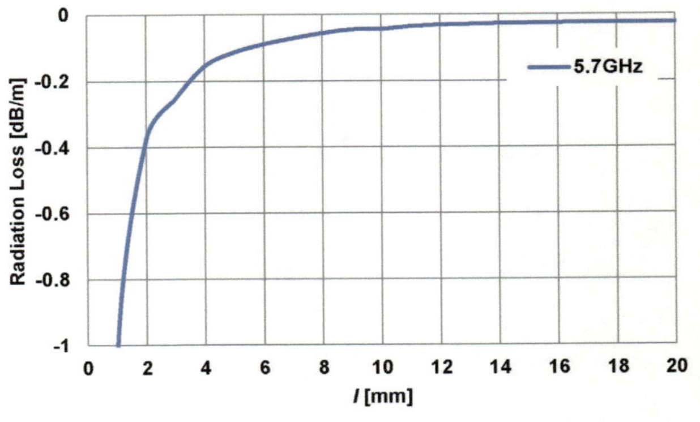

Figure 17.
Relationship between slit length and radiation.

$l > 12.0$ mm. If the open end of the slit is viewed as an ideal opening, when $l = \lambda/4$, the slit is short-circuited from inside the tube. This means that the slit can function as a leakage prevention structure.

However, t is small. It is necessary to consider that capacitance also exists between the slits. From these results, it is confirmed that, to obtain low-leakage characteristics, t should be shortened and $b \fallingdotseq \lambda/4$, so it is sufficient to adjust $l \fallingdotseq b$.

4. Evaluation of the communication experiment

The leaky waveguides were cascaded using joints. In this study, a non-leaky waveguide was used for the joints and the design was modified as shown in **Figure 18**. The insertion length of the leaky waveguide was 14 mm ($\fallingdotseq \lambda/4$). As shown in **Figure 19**, when a 14 mm length is inserted, a short-circuit boundary is provided in the inserted portion, which means that leakage at the joint portion is reduced. In this experiment, a leaky waveguide of 1 m length was used, and the joints were placed by alternately combining them.

Figure 20 shows the configuration of the experiment. The access point (e.g., ACERA 850F, FURUNO) was connected to the leaky waveguide via a coaxial waveguide conversion connector, and the leaky wave from the waveguide was received by the receiving antenna. At this point, the orientation of the receiving antenna from the waveguide was set as the x-axis, the antenna height direction was set as the y-axis, and the installed direction of the waveguide was set as the z-axis. The end of the waveguide was connected to a terminating resistor or to a spectrum analyzer via a connector.

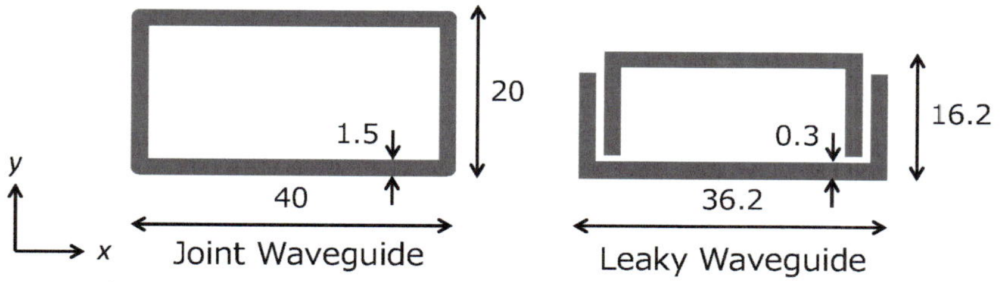

Figure 18.
Cross-sectional dimensions of the joints and leaky waveguides.

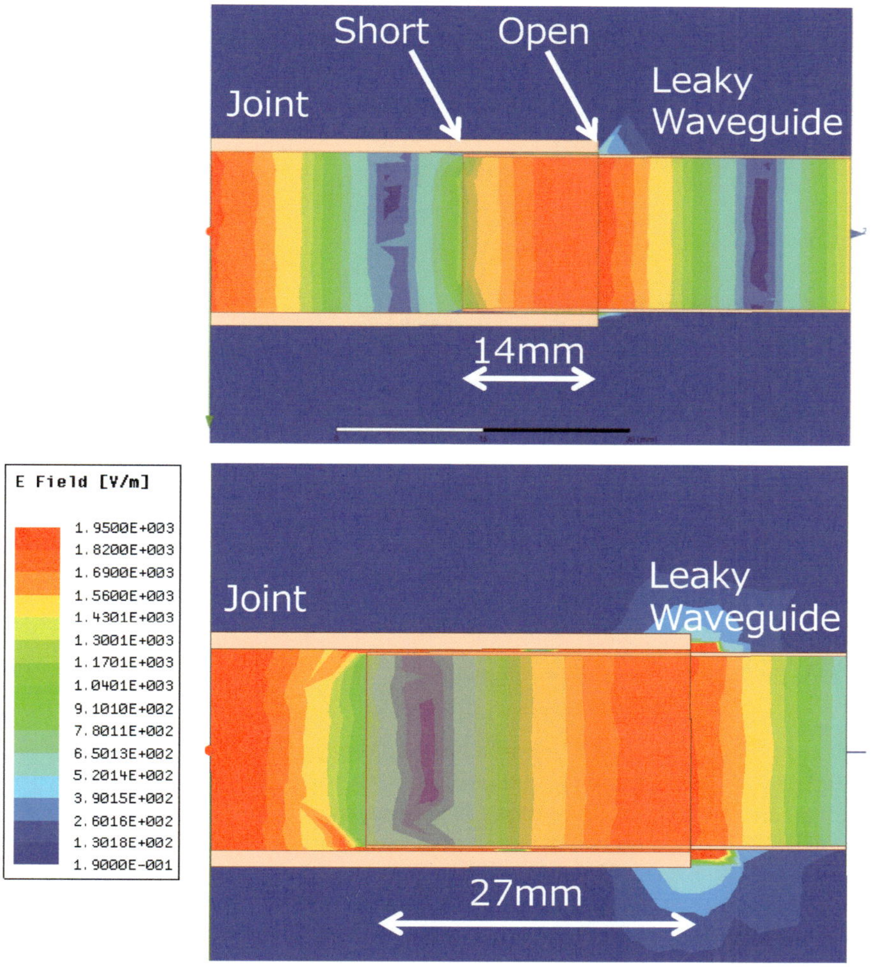

Figure 19.
Radiation from the joint (electric field).

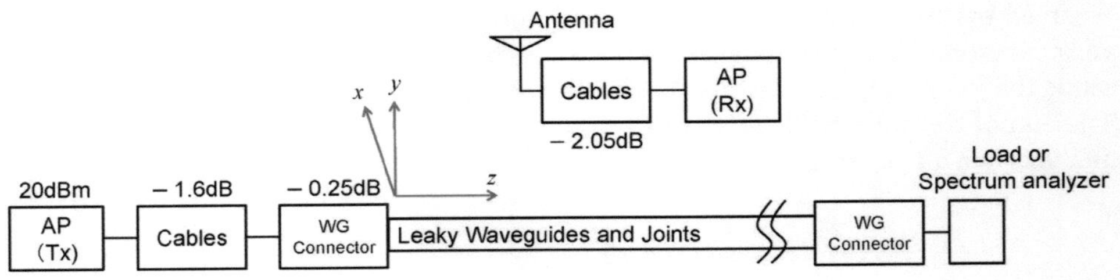

Figure 20.
Experimental configuration.

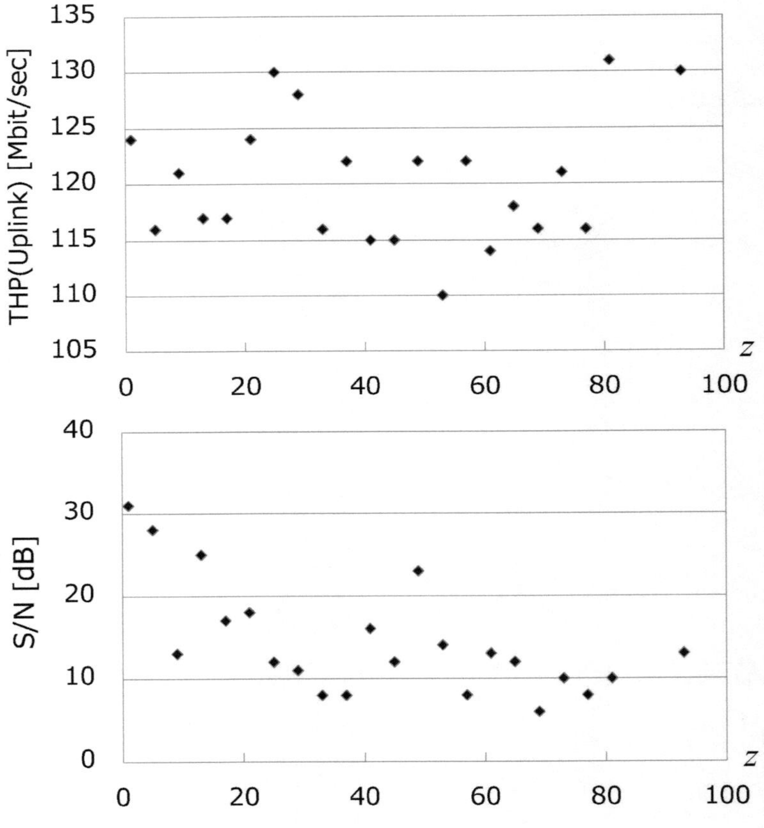

Figure 21.
*Throughput and S/N evaluation results (*x = 300 mm *and* y = 400 mm*).*

As a result of laying the waveguide over 120 m and measuring the power with the spectrum analyzer located at the end, it was found that the loss was 35.8 dB, i.e., approximately 0.32 dB/m. The output frequency was 5.6 GHz and the bandwidth was 40 MHz. **Figure 21** shows the evaluated throughputs and S/N ratios for the conditions x = 300 mm and y = 400 mm. In the circumference of the waveguide, good communications were possible up to 93 m. **Table 2** shows the results at z = 21 m and **Table 3** shows the results at z = 81 m. A S/N ≥ 5 dB was the approximate index for communications, and this index provided good communications.

S/N [dB]

y/x	0	150	300	450 mm
1000	18	22	19	20
800	15	22	21	17
600	–	22	18	11
400	–	26	26	23

THP(Uplink) [Mbit/sec]

y/x	0	150	300	450 mm
1000	117	119	122	118
800	116	117	117	120
600	–	117	117	117
400	–	118	122	116

Table 2.
Evaluation results at z = 21 m.

S/N [dB]

y/x	0	150	300 mm
900	10	10	11
800	14	13	14
700	8	11	13
600	10	12	8
500	14	12	10
400	12	17	7
350	13	9	6

Table 3.
Evaluation results at z = 81 m.

5. Conclusion

In this research, we proposed a low-leakage dual-plate waveguide and demonstrated the results of evaluating long-distance communications using this waveguide. As a result of this evaluation, under the conditions of $x = 300$ mm and $y = 400$ mm, it was shown that communication was possible at speeds of 100 Mbps or more up to a distance of 92 m. We will continue to consider using this system in tunnels.

Acknowledgements

I am grateful to Dr. Patrick Steglich of the book editor who gave me the opportunity to write this chapter. We thank Mr. Nobuyuki Watanabe and Mr. Toshifumi Sakai (FURUNO) for their help in this study. Moreover, I respect the great achievements of my predecessors whose studies I have cited.

Author details

Shotaro Ishino[1*], Satoshi Denno[2], Narumi Yashiro[2] and Seiichi Suzuki[2]

1 Furuno, Hyogo, Japan

2 Toda Corporation, Tokyo, Japan

*Address all correspondence to: shotaroh.ishino.qx@furuno.co.jp

References

[1] Li J, Zhao Y, Zhang J, Jiang R, Tao C, Tan Z. Radio channel measurements and analysis at 2.4/5 GHz in subway tunnels, China. Communications. 2015; **12**(1):36-45

[2] Guan K, Ai B, Zhong ZD, et al. Measurements and analysis of large-scale fading characteristics in curved subway tunnels at 920 MHz, 2400 MHz, and 5705 MHz. IEEE Transactions on Intelligent Transportation Systems. 2015;**16**(5):2393-2405

[3] Guan K, Zhong ZD, Ai B, et al. Complete propagation model in tunnels. IEEE Antennas & Wireless Propagation Letters. 2013;**12**:741-744

[4] Guan K, Zhong ZD, Alonso JI, Rodriguez CB. Measurement of distributed antenna systems at 2.4 GHz in a realistic subway tunnel environment. IEEE Transactions on Vehicle Technology. 2011;**61**(2):834-837

[5] Wang JH, Mei KK. Theory and analysis of leaky coaxial cables with periodic slots. Institute of Electrical and Electronics Engineers. 2001;**49**(12): 1723-1732

[6] Morgan SP. Prediction of indoor wireless coverage by leaky coaxial cable using ray tracing. IEEE Transactions on Vehicular Technology. 1999;**48**(6): 2005-2014

[7] Ishino S. Novel waveguide technologies and its future system applications, surface waves. In: New Trends and Developments. Croatia: IntechOpen; 2017. pp. 1-15

[8] Goldstone L, Oliner A. Leaky-wave antennas I: Rectangular waveguides. IRE Transactions on Antennas and Propagation. Oct. 1959;**7**(4):307-319

[9] Menzel W. A new travelling wave antenna in microstrip. In: 1978 8th European Microwave Conference. September 1978

[10] Ishino S, Etani S, Matsumoto S. Study of a 5 GHz dual-plate leakage waveguide for wireless LAN infrastructure construction. IEICE Tech. Rep. AP2018–37. 2018;**118**(103):1-4 [Japanese]

Section 4

Guided Waves in Free Space

Chapter 7

Study of Refraction Effects for Propagation over Terrain

Vladimir Schejbal, Ondrej Fiser and Vadim Zavodny

Abstract

This chapter investigates the radio-wave propagation above irregular ground, including the troposphere, using physical optics computation. We briefly describe used simplifications, which substantially reduce numerical simulations. Using the principle of stationary phase, we can approximate the propagation over a terrain (the PO approximation of the vector problem with a 3-D surface) with a 2-D surface. Moreover, we approximate the reflection coefficient for a surface with random deviations considering the surface standard deviation and the local Fresnel reflection coefficient for the smooth ground. We present the novel computations of physical optics for investigations of radar coverage diagrams. We consider both monostatic and bistatic radars, the far-field antenna measuring ranges, and studies of air refraction index. We validate the calculations by both experimental results and the other numerical simulations. The experimental results changed during seasons and according to terrain and troposphere conditions including vegetation, cultivation, snow, and air temperature and pressure.

Keywords: electromagnetic propagation, electromagnetic reflection, microwave propagation, losses, radar antennas, electromagnetic refraction

1. Introduction

Propagation of radio waves above earth is very challenging for uncountable communication tasks comprising the radar coverage and far-field antenna measurement ranges. Several methods have been described [1–13] such as geometrical optics (GO) and various modifications of the geometrical theory of diffraction. Full wave methods are rather demanding, bearing in mind the memory and central processing unit (CPU).

We present a brief description of the physical optics (PO) method [14–22] applied for irregular ground reflection considering both horizontal and vertical polarizations, electrical properties of earth (i.e., reflection coefficient), scattering of radio waves from random surfaces, and the shadow radiation. The improved computation of radio waves above uneven ground uses PO and line integrals, taking into consideration the vector problem and shadowing [18, 19]. This is a more consistent method for low-altitude fields and diffraction zones without any additional alternatives.

We perform new numerical simulations, which we compare with ample experimental results and other numerical simulations such as the parabolic equation method (PEM) for altered environment circumstances and modifications.

The evolutionary processes are also discussed and are fully referenced below. We investigate radar coverage diagrams including monostatic and bistatic radars, far-field measuring range of antenna with separation roughly 1 km, and troposphere refraction effects for separation of 49.8 km. The longstanding testing demonstrates that the PO could present reliable computations for low heights and diffraction zones for numerous irregular grounds and real distribution of refraction.

2. Computation of irregular ground reflection

Based on the PO method [3, 4], the computation of irregular ground reflection was derived [14]. However in the 1970s, it was necessary to diminish memory and CPU time. Therefore using a stationary phase method, the line integrals were computed instead of surface integrals, and scalar solutions were only employed. The earliest method has been progressively enlarged. Gradually, we have included various options considering both horizontal and vertical polarizations, electrical properties of earth (i.e., reflection coefficient), the scattering of radio waves from random surfaces, and the shadow radiation. Considering Ufimtsev's results [23–25] a totally new approach to analyze propagation over irregular terrain could be used. The improved computation of radio waves above uneven ground uses PO and line integrals and takes into consideration the vector problem and shadowing [18, 19]. That is a more consistent method for low-altitude fields and diffraction zones without any additional alternatives.

The PO method analyzes an antenna A above the ground as is displayed in **Figure 1**. The sum of incident, $\boldsymbol{E}_i(\mathrm{P})$, and scattered, $\boldsymbol{E}_s(\mathrm{P})$, electric fields could be used everywhere for calculation of the resultant field. The vector of the total electric field, $\boldsymbol{E}(\mathrm{P})$, at point P is

$$\boldsymbol{E}(\mathrm{P}) = \boldsymbol{E}_i(\mathrm{P}) + \boldsymbol{E}_s(\mathrm{P}). \tag{1}$$

The $\boldsymbol{E}_i(\mathrm{P})$ field may be calculated as a spherical wave. An actual scattered body is substituted by the corresponding currents induced on its surface; that is, an allocation of corresponding currents in free space should be computed, which transmit without restriction in all paths. If these currents were computed exactly, they would deliver the accurate scattering results.

Radio-wave scattering by certain impenetrable bodies with local reflection coefficients can be computed using PO [23–25]. According to **Figure 1** the earth's surface may be separated into illuminated, S_{il}, and shadowed, S_{sh}, parts with a shadow curve between them.

According to [3] the GO is a limiting form of the PO. The spread border of the shadow in the diffraction phenomena becomes the sharp shadow of GO as the

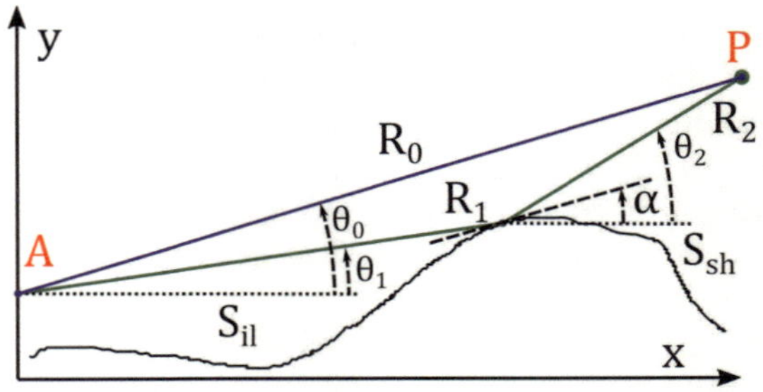

Figure 1.
Propagation geometry.

wavelength tends to zero. Therefore both GO and PO are very useful in the analyses of microwave propagation. Bearing in mind the PO, body surface fields are calculated using the GO. Therefore, the induced sources are only defined on the S_{il} of the scattering object. The components of the S_{sh} are set to zero.

The propagation above the earth (i.e., the PO calculation of the vector problem with 3-D surface) may be substantially reduced, both from memory and CPU time points of view. Using the principle of stationary phase [26]

$$\int_{-\infty}^{\infty} \exp\left(-j\pi z^2/2\right)dz = \sqrt{2}\exp\left(-j\pi/4\right) \tag{2}$$

where the horizontal polarization component $E_{sz}(\mathrm{P})$ and the maximum value of the incident electric vector E_0 at a distance R_0 are [14]

$$\frac{E_{sz}(P)}{|E_0|} = \frac{R_0 e^{j\pi/4}}{2\sqrt{\lambda}} \int_a^b f(\theta_1)\left[(1-\Gamma)\sin(\theta_1-\alpha) + (1+\Gamma)\times\sin(\theta_2-\alpha)\right] \frac{e^{-jk(R_1+R_2-R_0)}}{\sqrt{R_1R_2(R_1-R_2)}}\frac{dx}{\cos\alpha} \tag{3}$$

where R_0, R_1, R_2, θ_1, θ_2, and α are shown in **Figure 1**, $f(\theta_1)$ is the normalized antenna radiation pattern with phase center at point A at height h_A over the terrain, Γ is the Fresnel reflection coefficient (local reflection coefficient), $k = 2\pi/\lambda$, λ is the wavelength, and a, b are limits of the illuminated part S_{il}. A similar equation can be derived for vertical polarization using H_{sz} [15]. Therefore, this method takes into account the polarization. A rather analogous approach is used in [9].

The reflection coefficient for a surface with random deviations could be approximated by

$$\Gamma = \Gamma_0 \exp.\left[-2(2\pi\sigma \sin\gamma_0/\lambda)^2\right] \tag{4}$$

where σ is the surface standard deviation, Γ_0 is the local Fresnel reflection coefficient for the smooth surface for the horizontal (or vertical) polarization, and γ_0 is the grazing (reflection) angle—the angle between the tangent and the incident (reflected) ray. Obviously, better models were proposed [4, 5, 11] considering surfaces as random processes. However, they are rather complicated, and selection of parameters such as correlation length could be questionable.

The improved computations [18, 19] of scattered fields consider the reflected radiation component, E_{sz}^{ref}, (with the reflection coefficient Γ terms) and the shadow radiation component, E_{sz}^{sh}:

$$S_{sz}^{ref}(P) = \frac{|E_0|R_0 e^{j\pi/4}}{2\sqrt{\lambda}} \int_a^b f(\theta_1)\Gamma[\sin(\theta_2-\alpha) - \sin(\theta_1-\alpha)] \times \frac{e^{-jk(R_1+R_2-R_0)}}{\sqrt{R_1R_2(R_1+R_2)}}\frac{dx}{\cos\alpha} \tag{5}$$

$$S_{sz}^{sh}(P) = \frac{|E_0|R_0 e^{j\pi/4}}{2\sqrt{\lambda}} \int_a^b f(\theta_1)[\sin(\theta_2-\alpha) + \sin(\theta_1-\alpha)] \times \frac{e^{-jk(R_1+R_2-R_0)}}{\sqrt{R_1R_2(R_1+R_2)}}\frac{dx}{\cos\alpha} \tag{6}$$

Similar equations may be obtained for vertical polarization utilizing the H magnetic field. Thus, the PO method respects the polarization.

The reflected component depends on the local reflection coefficient, Γ, considering the surface standard deviation, σ. On the other hand, the shadow-emitted power corresponds to the entire power incident on a scattered body, and it is not determined by the reflection coefficients. Considering the shadow contour theorem, it is not determined by the structure of the scattered body and is completely influenced only by the dimensions and the shape of the shadow line. For the shadow sector, at a limited extent from the scattered body (at the rear of body), the shadow emission for a very short wavelength can be counted as an electromagnetic wave beam. This beam approximately cancels the incident field, and the reflected beams nearly disappear. The shadow radiation gives origin to edge waves, creeping waves, and surface diffracted rays.

The previous computations of low-altitude propagation (LAP) and transient zone (TZ) [16] use the knife-edge diffraction [1] and Fock's spherical surface solution [5]. Both are scalar solutions, which neglect terrain imperfections. These methods are well known and have been extensively used. They have been proven as very efficient approximate methods for real terrain both analytically and experimentally. The transient zone would be considered, if the differences between the reflected and incident rays were less than a third of the wavelength. The low altitude would be considered, if the differences were less than $\lambda/2\pi$. Therefore the previous computations are relatively artificial as two quite different approximations are used for the computation, and the transient zone limits are only supported by an ad hoc assumption.

The comparison [18, 19] of the previous and new methods shows that the improved analysis of propagation over irregular terrain could be much more useful and accurate. Thus, Eqs. (5) and (6) may be employed for computation of both illuminated and shadow radiation. The calculation may be performed for greater heights (greater differentiations between incident and reflecting beams) together with lower heights (i.e., it is not required to compute the low heights and some transient regions). This offers much more consistent results, which consider the polarization even for the shadow zone.

A beam spreading via the lower troposphere refracts according to the refraction index gradient. As the refraction indexes change primarily with height, only the gradient of the vertical refraction index, n, is generally respected. If the refractivity height profile is linear, i.e., the refraction gradient is stable along the ray trajectory, then the transformation [3] considering a hypothetical Earth of effective radius R_e and linear ray trajectories can be used.

For calculation, the piecewise approximation of the surface is used in **Figure 2a**. The length of arc, r_B, and height, h_B, as shown in **Figure 2b**, are read. The coordinates (x_B, y_B) are provided by the subsequent equations [16]:

$$\begin{aligned} x_B &= (R_e + h_B)\sin\alpha_z \\ y_B &= (R_e + h_B)\cos\alpha_z - R_e \\ \alpha_z &= r_B/R_e. \end{aligned} \tag{7}$$

Obviously, an effect of these equations is greater when r_B and/or h_B are greater. This is demonstrated by examples in the next sections.

For a fast oscillating integrand, when the phase variation exceeds many times 2π, the utilization of usual trapezoidal or Simpson's quadrature formulas is inadequate. However, the numerical simulations (Eqs. (5) and (6)) may be performed by the generalized trapezoidal method [20], which is very efficient considering memory and CPU time. This method is based on the piecewise approximations of amplitudes as well as phases with equal intervals without any significant limitations. Numerical integrations of both real and imaginary parts are performed by usual trapezoidal method. It was found by numerous computations for various kinds of terrains that for frequencies less than 30 GHz, the integration steps may be 5–10 m [21].

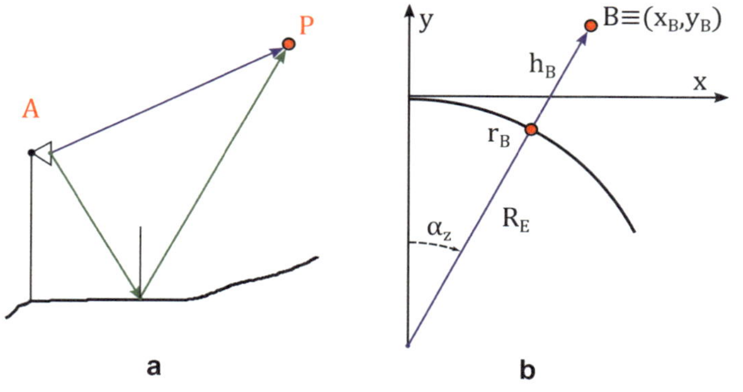

Figure 2.
(a) Earth surface and (b) curvature correction.

Actually, the usual requirement [9] that the spatial sampling resolution is less than λ/2 (1.5 cm for 10 GHz) is created mostly by aliasing.

However, for both analyses and syntheses, the simplified computation of the electrical field above an uneven earth [22], derived from above described method, could be used. This simplified method could compute Eqs. (5) and (6) using suitable simple approximations of Fresnel integrals. Clearly, the most important portions of piecewise approximations are in the vicinity of stationary phase points (i.e., points, where rays reflected from surfaces appear to come from mirror images). Therefore, the simplified method creates a more precise method than GO methods.

The difficulties of the described procedure are created by calculations used for the ground field. They may be reduced by using the physical theory of diffraction (PTD) [13, 24], which is a substantial expansion of PO. Furthermore, the novel variety of PTD [25] is acceptable for all scattering paths, particularly those that may contain forward scattering.

The normalized resulting field

$$A = 20 \ \log|E(P)/E0|, \tag{8}$$

where $E(P)$ is the resulting field at P and $E0$ is the maximum incident electric field, which is determined for comparison of numerical simulations with experimental and other method data.

3. Radar coverage

A radar coverage diagram [3, 15, 26–30] comprises a volume inside in which the field is greater than the minimum useful value. The PEM models are very beautiful, but they ask for larger memories and CPU times, particularly for higher frequencies, elevation angles, and long ranges. Therefore diverse hybrid models have been produced by joining different models such as PEM and GO.

In the bistatic radar, unlike the monostatic radar, the transmitter and receiver are separated by a distance comparable to the target-to-receiver range [27]. The radar equation states

$$P_R = P_T G_T G_R \frac{f_T^2(\Phi,\Theta) f_R^2(\Phi,\Theta)\lambda^2\sigma_B}{(4\pi)^3 R_T^2 R_R^2 L_{TR}}, \tag{9}$$

where P_T, P_R are transmitted and received powers, G_T, G_R are transmitter and receiver antenna gains, f_T, f_R are transmitter and receiver antenna characteristics, λ is the wavelength of transmitted signal, σ_B is a radar cross section (RCS), and L_{TR} is loss on the transmitter-receiver path.

3.1 Monostatic radars

Usually vertical coverage diagrams are shown for free-space and plane ground effects, which consider a standard effective radius of $R_e = 8.5 \times 10^6$ m such as shown in **Figure 3**. This offers very useful qualitative ideas.

However, this approach is not quite satisfactory. Therefore, coverage diagrams of manufactured monostatic radars, which usually employ the same antenna for both transmitting and receiving antenna (i.e., $G_T = G_R$ and $f_T = f_R$), have been analyzed for numerous airports and radars, using the described PO method since the 1970s [31].

Figure 4 shows the new numerical simulations of electric fields for elevation angles, θ. The free-space and PO calculations of ground impacts for plane approximation of airport terrain profile as well as two effective radiuses (standard $R_e = 8.5 \times 10^6$ m and $R_e = 10.2 \times 10^6$ m) are compared. Even if it seems that these effective radiuses give very similar results, we should consider that the constant height contours of the coverage diagram curve downward as is shown in **Figure 3**.

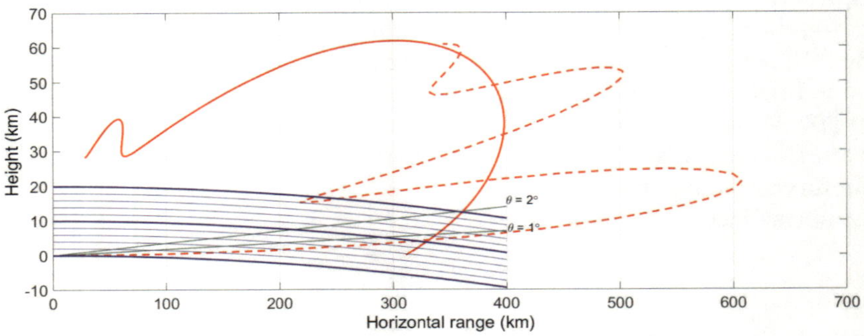

Figure 3.
Vertical coverage diagrams of radar for free-space (solid line) and ground effects.

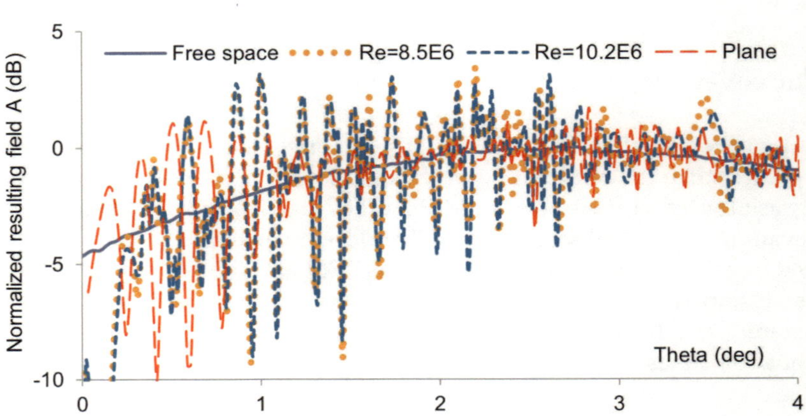

Figure 4.
Numerical simulations for free-space and PO calculations of ground impacts for plane approximation of airport terrain profile and two effective radiuses.

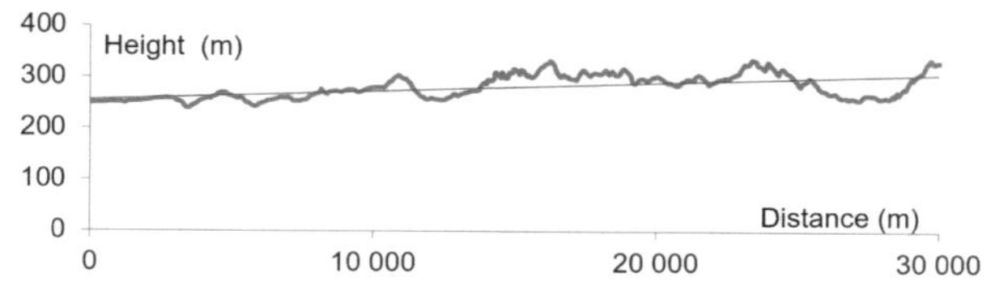

Figure 5.
Airport terrain profile with straight line fitting.

These simulations are performed for the novel radar development at an airport with terrain profile of a specified azimuth, displayed in **Figure 5**. We considered the surface with relative permittivity of ε_r = 2.9–0.044j and standard deviation of σ = 0 m for any part of the terrain.

However, this selection is not very important according to the detailed analyses [14, 15]. Obviously, a reflection-interference lobe pattern can be clearly seen. We used a straight line fitting shown in **Figure 5** for comparison. However, numerical simulations of plane approximation shown in **Figure 4** demonstrate a mere quality agreement, i.e., the plane approximation could be only used for a very rough calculation. This is due to fact that least squares data fittings, which seem very reasonable from mathematical point of view, use artificial slopes of terrain, especially for particular parts of surfaces.

On the other hand, calculations using the partial wave method [15, 32, 33] and mentioned simplified computations [22] correspond to PO simulations, even if we use a fit by eye for the set of data, because slopes of terrain correspond to piecewise approximations.

Obviously, the change of effective radius, R_e, is not usually substantial, especially for lower elevations. The used code for PO approximations allows only the utilization of one constant effective earth radius, R_e, for variable altitudes. This cannot be used for greater heights. However, the calculations of radar coverage could be more accurate using the recommendations ITU-R [34, 35] for the computations of refraction effects (estimation of the apparent elevation angle).

Experimental validations of vertical coverage calculations of radars are rather difficult as the RCS of various targets is usually extremely variable (obviously except conducting sphere). Usually, customers ask for confirmations of the radar coverage diagrams by test flights such as shown in **Figure 6**. However, that depends on several items such as air refraction effects and the RCS of the operated airplane, which are extremely changeable [27, 28, 36]. The RCS varies as a function of aspect angle and frequency (the period of the variability changes from seconds to a few tenths of a second). However, thanks to plentiful test flights made at different airports for changeable azimuths (therefore completely dissimilar topography profiles) and thorough analyses, we could state that PO simulations correspond to test flights.

The effective elevation pattern clearly depends on superposition of the direct propagating signal with reflecting signals. However, according to skills with operations and testing of radars nearby to airports with grassy vegetation, the diminishing of the reflection coefficient for angles up to 2° is not important [32]. We can conclude that the radar coverage diagram of certain specific monostatic radar could be very useful as neighboring terrains of radar sites are usually very similar, and therefore it could be considered as a typical case.

3.2 Bistatic radars

There are several competitive signals reducing the maximum range and complicating the signal processing. They are the direct signal, correlated reflections of

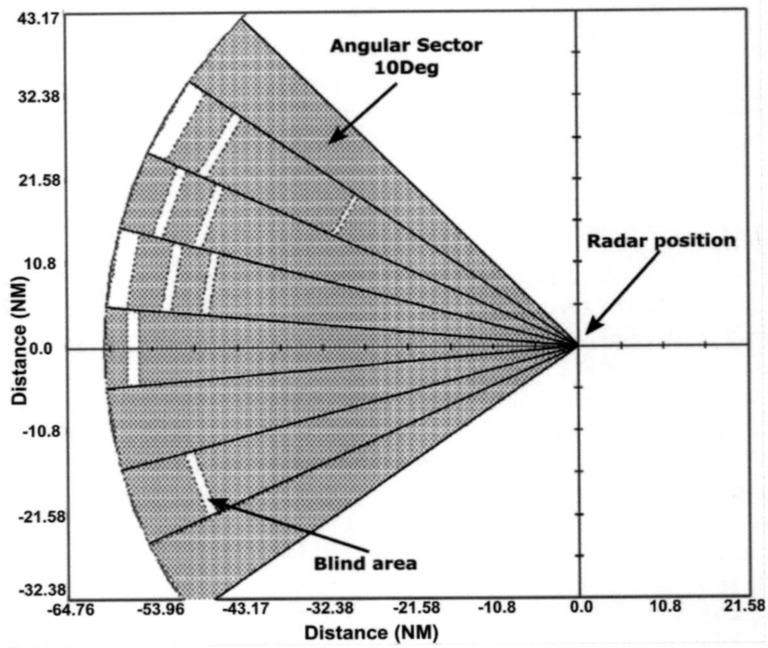

Figure 6.
Test flight.

terrain objects (clutter), non-correlated signals transmitted by other sources at the same frequency (electromagnetic noise), and the thermal noise. Clearly, the investigation of bistatic radars and especially passive coherent locator (PCL), which uses transmitters of opportunity such as FM radio broadcasting, is very complicated [27, 37, 38].

As the transmitter and receiver are separated, the ground effects should be analyzed independently. That means, the propagation above the terrain between transmitter(s) and target and receiver(s) and target must be examined. This case is much more complicated than monostatic radar. However, **Figure 3** could be very useful for investigation of bistatic radars for various different situations. Moreover, PO modeling very easily permits the typical configuration calculations for transmitters of opportunity.

Obviously, the bistatic RCS should be also investigated. Usually, a bistatic RCS is lower than the monostatic RCS measured on the bisector of complicated targets. However, some target aspect angles can generate a low monostatic RCS and high bistatic specular RCS. A limiting case of the bistatic geometry occurs when the target is on the transmitter-receiver baseline. A considerable improvement of scattering can be generated, as the forward scatter RCS, σ_f, is approximately

$$\sigma_f = 4\pi S^2/\lambda \tag{10}$$

where S is the target projected area and λ is a transmitted wavelength even for stealth targets with ideally absorbing surfaces [27].

4. Antenna far-field measuring ranges

We measure an antenna under test (AUT) on an antenna range [39–41]. The plane wave of uniform amplitude and phase is an ideal situation for measuring of far-field electromagnetic wave features of the AUT. This wonderful situation is not realizable. However, it can be roughly made, if we use a large separation between the AUT and the transmitting (or receiving) antenna at an outdoor range. When we enlarge the separation, the spherical phase-front becomes more planar over the AUT aperture. When the separation equals $2D^2/\lambda$, where D is the largest diameter of the AUT, then the maximum phase error is about 22.5°.

Furthermore, reflections from the ground and adjacent objects could erroneously affect the AUT illuminations. Apparently, the electromagnetic field quality of the quiet zone depends on AUT features. Thus, numerical simulation analyses are very motivating. Applicable suppression of reflected signals should be done by a combination of line-of-sight clearance, transmitter and/or receiver antenna directivities, and sidelobe suppression. Perhaps, range ground screening could also be considered.

The proposed PO method [18] analyzes the illuminated part of terrain and abrupt change of height, which create the shadowed part. Naturally, the utilization of knife-edge diffraction and Fock's spherical surface solutions, which neglect terrain imperfections, creates only approximate solutions. However, the comparison shows that the PO method produces acceptable results, as this method actually replaces a real scattering object by the equivalent currents. Moreover, problems may be reduced by using the physical theory of diffraction [24, 25].

Obviously, a smaller AUT asks for a smaller quiet zone, but spurious signal suppressions by the AUT could be very poor. On the contrary, a larger AUT requires a larger quiet zone, but spurious signal suppressions by the AUT could be much improved.

We could test the function of any antenna through concrete solid angles and frequency bands. The antenna features are typically quantified by the requests of operating systems and describe areas where they are significant. Plentiful measurements have been obtained thanks to thorough tests of the far-field ranges for different situations since the early 1970s (as for any original antenna type, the vertical range illumination was tested). Obviously, any discrepancies and changes have been comprehensively analyzed to discover the possible reasons of those effects.

Of course, the effect of terrain on scattering field is very important. But, the numerous technical and economic issues such as possibilities of electrical power supply, roads suitable for tested antenna transport, effect of nearby objects (buildings, woods), electromagnetic interference sources, expenses, and total budget should be considered. Therefore, various sites have been thoroughly analyzed. In fact, the program [14] was proposed for different studies of projected and/or built far-field ranges.

Therefore, we present the novel comparison with different relative permittivity values ε_r and standard deviations σ in **Figure 7** for the profile shown in **Figure 8**. The h = 0 height corresponds to upper positioner placed on the tower.

The simulations of dry ground of ε_r = 3.2–0.015j and wet ground of ε_r = 30–2.5j are shown. It can be seen that the results are nearly the same. Experimental values and numerical simulations were performed with a transmitting reflector diameter of D = 3 m. To validate the influence of greater beam width, the computations with the reflector diameter of D = 0.6 m (ε_r = 30–0.02j) are also presented. Clearly, standard deviations σ of random surface deviations are more significant, especially for larger beam width (i.e., D = 0.6 m).

The problems of ground scattering could be diminished by using fences, which could be very demanding [40]. The diffraction fences on the range can reduce the level of reflections. However, the fences inevitably introduce disturbances in the incident field due to diffraction effects along their wedges. Therefore, a practical

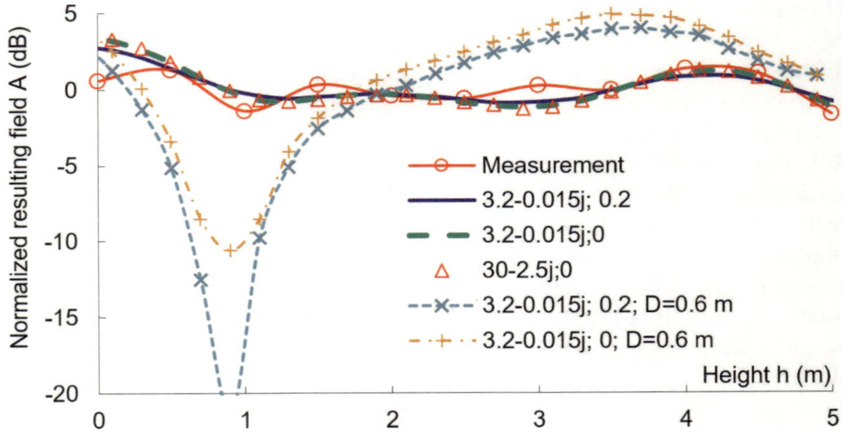

Figure 7.
Measurement, numerical simulations for ε_r = 30–2.5j and ε_r = 3.2–0.015j, ground standard deviations of σ = 0.2 m and σ = 0 m, and calculations with reflector diameter of D = 0.6 m for ε_r = 3.2–0.015j, and ground standard deviations of σ = 0.2 m and σ = 0 m.

fence proposal is a compromise between reflection reductions and residual diffraction. Of course, it has been shown [18] that Eqs. (5) and (6) may be utilized for computation of both illuminated and shadow radiation considering the polarizations even for the shadow zone. That is numerical simulations may be performed for greater as well as lower heights without any artificial combinations of different methods. Considering the complex Fresnel integral [1] and Eq. (2), the Cornu spiral derived from straight-edge diffraction can be applied. Therefore, the length of reflected rays shown in **Figure 8** could be very useful for fence design.

Wedge diffractions of fences could be diminished using tuned slots (which are effective at a single frequency but are frequency sensitive) or serrations along the edge. Both approaches ask for structures which expand several wavelengths above the edge. The most understandable design is to maintain low-level illumination of the fence edge, if possible. That sometimes leads to the low multiple-fence design with proper tilting. However, usually one to three fences are used.

Generally, the differences between measuring, when the probe (small horn antenna) is moved between h = 0 and h = 5 m, and calculations may be partially clarified considering reflective coefficient variations and scattering from objects, which are nearby the positioner. Such objects include a tower construction and safeguard bars.

The reflective coefficients fluctuate due to seasonal ground circumstances as the ground may be overgrown by plants, coated by snow or farmed. They influence both scattered and resulting fields. They are not frequently significant for low

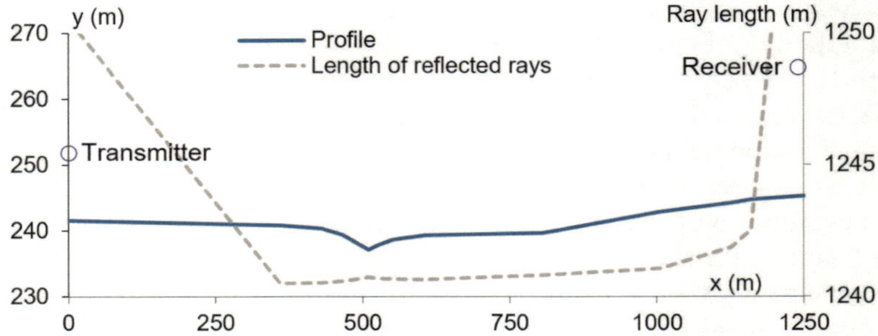

Figure 8.
Profile of far-field range and the length of reflected rays.

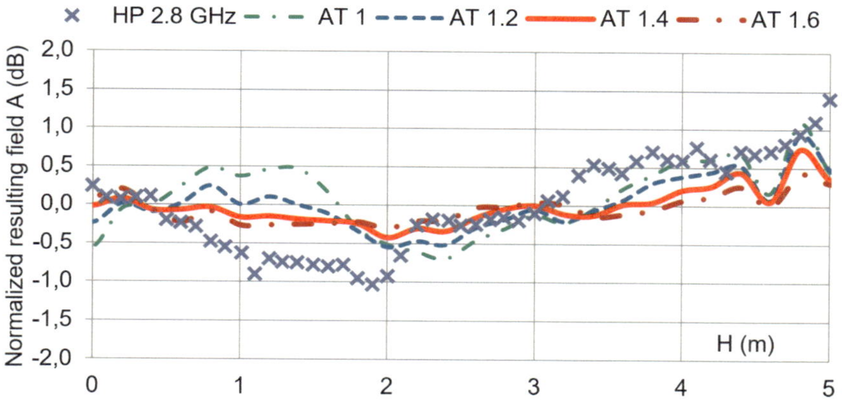

Figure 9.
The comparison of measurements with antenna tilt 1.4° (HP 2.8 GHz) and calculations of normalized resulting field for frequency of 2.8 GHz and the antenna tilt of 1° (AT 1), 1.2° (AT 1.2), 1.4° (AT 1.4), and 1.6° (AT 1.6).

random deviations, as the local reflections are nearly equal to −1 for low grazing angles irrespective of polarization. However, **Figure 7** shows that larger ground deviations may influence the measured values more significantly. Site tests and numerous computations since the 1970s validate these data. Moreover, the experiments and computations have been done with various ranges with quite different terrain profiles. Various initial and reconstructed towers, different transmitter antennas (such as prime focus and dual-reflector Cassegrain antennas with smooth-wall or conical corrugated horns), and several types of probes are used for both linear and elliptical polarizations and frequency bands.

Actually, the results fluctuate during seasons and due to location of auxiliary equipment (including occasionally crane). The experiments show that reflection coefficients of antenna ranges are diminished in summer, when a terrain is covered by wheat or other vegetations for frequency bands of 1 up to 10 GHz.

Numerical simulations have been used both for design and optimizing of the rebuilt antenna range [14–18, 36, 39, 42]. **Figure 9** shows the comparison of measurements with antenna tilt 1.4° (horizontal polarization HP 2.8 GHz) and calculations of normalized resulting field A (dB), using the horizontal polarization and frequency of 2.8 GHz. Various antenna tilts of 1° (AT 1), 1.2° (AT 1.2), 1.4° (AT 1.4), and 1.6° (AT 1.6) are analyzed. It is quite clear that experimental optimizing, when we consider the antenna tilt as well as frequencies, polarizations, and random deviations, is very time-consuming. Therefore, numerical simulations are extremely useful. They could discover quite new phenomena and create new concepts.

Obviously, numerical simulations have been analyzed considering disturbing effects of various obstacles such as nearby constructions, trees, and changeable surface. This is important especially for bigger random deviations, i.e., the surface standard deviation, σ.

5. Effects of air refraction index

Electromagnetic wave propagation in the troposphere varies according to the air refractive index [3, 5, 34, 35]. The various experimental analyses of radio-wave characteristics and atmospheric refracting "*N* unit" layers have been already published such as [43–46]. All these factors can be used for calculation. The enduring testing of physical tropospheric features were made at the receiving site on a tower with 19 different elevations from 5 m up to 147 m with mean extent spans of about 7.5 m. Simultaneously, the five receiving 0.65 m dishes at several elevations measured the electromagnetic field in the troposphere with the link length of R_0 = 49.8 km. Bearing in mind the 1-day measuring of refracting N units, **Figures 10** and **11** show the comparison of measurements and PO computations.

Figure 10 shows the spreading of measured values during the same day denoted by MEAS.MAX and MEAS.MIN, which show measured maxima and minima. Comparisons of maxima and minima of measurements during a day with PO numerical simulations of 0.65 m diameter antenna, surface with ε = 15–3.5j, σ equals to 0 or 0.2 m, and standard $R_e = 8.5 \times 10^6$ m reveal that the simulations using only one R_e cannot be used for a fine modeling. Even if we use $R_e = 6.9 \times 10^6$ m, which correspond to height of 51 m, the problem is not resolved for several heights. Therefore the other approach should be used.

The observation of refractive index distributions, for various heights during the same day, reveals that distributions are very changeable. Therefore, we cannot speak correctly about stratum formulation. Nevertheless, we select the same R_e effective radiuses for individual "layers" for 11 AM. The heights of 51, 61, 90, 120, and 145 m correspond to PO simulations with individual R_e selections for each height ("layer"). Therefore the small parts of graphs are only displayed in **Figure 11**.

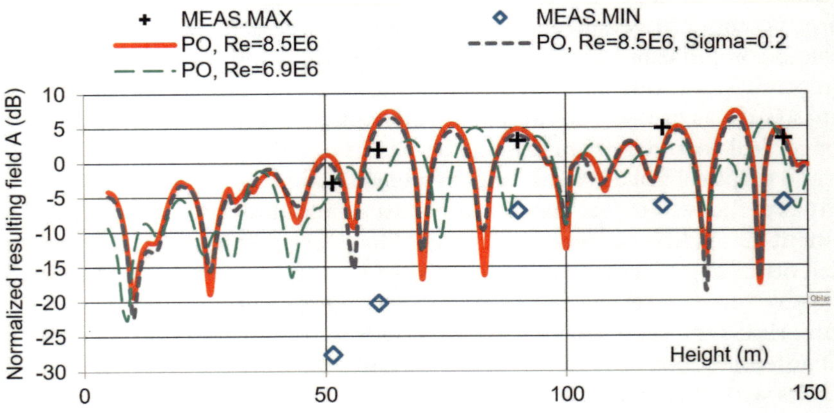

Figure 10.
Comparison of measurements, where MEAS.MAX and MEAS.MIN are maxima and minima during a day, with the 0.65 m diameter and standard $R_e = 8.5 \times 10^6$ m (with σ equals to 0 or 0.2 m) and $R_e = 6.9 \times 10^6$ m for PO numerical simulations.

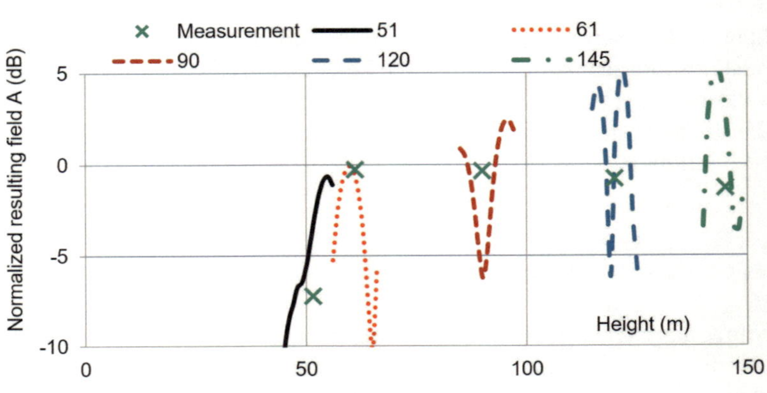

Figure 11.
Comparison of measurements with 0.65 m diameter for PO numerical simulations for heights of 51, 61, 90, 120, and 145 m.

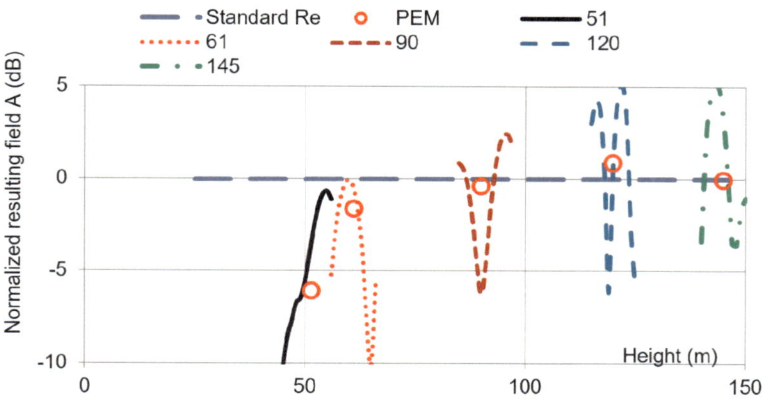

Figure 12.
Comparison of the standard R_e (without ground reflections), PEM, and PO numerical simulations for heights of 51, 61, 90, 120, and 145 m.

This demonstrates that the described PO method may offer reliable calculations of three-dimensional spreading of refraction.

Considering the 1-day measuring of refracting N units, **Figure 12** shows the comparison of computations using PO and PEM for 11 AM (i.e., the same time as in **Figure 11**). It can be seen that PO and PEM provide similar results, if we select the same R_e effective radiuses for individual "layers" (both for PEM and PO as correspond to the "measurement" values). The standard R_e shows calculations without ground reflections. Clearly, the effects of the antenna radiation pattern are negligible for low elevations.

The PEM has been expansively examined. Evaluation of experiments and numerical simulations using the PEM revealed that PEM simulations correspond mostly to measurements when a particular vertical gradient may be used. However, a simulation of multipath spreading for extremely confused circumstances was unacceptable. Unfortunately, the requirement that not less than three frequencies should be used at the same time to offer an obvious correspondence with theory [3] has not been accomplished as frequency of 10.671 GHz was only used. Using the PO method, the influence of refracting indexes was examined infrequently, and initial situations have been only announced [4, 6]. Therefore, new comparisons are performed.

However, the employed code for PO approximations allows only the utilization of one constant effective earth radius, R_e, for variable altitudes.

Using the code [18, 19] we could use different electrical parameters for any ground fragment. In spite of this, the detailed characteristics of the ground are not identified. In fact, they are not stationary and may change very rapidly. Thus, ε_r = 15–3.5j is only employed for computations of air refraction index influences.

6. Conclusions

This chapter briefly describes the PO approximation, which is frequently utilized. The novel comparisons using recent PO method are presented for measurements and numerical simulations. We have investigated experimentally different ground situations and variations of plants, snow, winter, or summer through plentiful years for different frequency bands and polarizations.

We analyze radar coverage diagrams considering the usual monostatic radars as well as bistatic radars. We can conclude that the radar coverage diagram of certain specific monostatic radar could be very useful as neighboring terrains of radar sites are usually very similar. Therefore, it could be considered as a typical case. Analyses of bistatic radars are difficult as the propagation above the terrain between transmitter(s) and target and receiver and target must be analyzed. This case is much more complicated comparing with monostatic radar. However, **Figure 3** could be very useful for investigation of bistatic radars for various different situations. Moreover, considering passive coherent locators, the PO modeling permits typical configuration calculations for various transmitters of opportunity.

The measuring antenna far-field range (relatively short distance about 1 km) shows generally that the differences between measuring, when the probe (small horn antenna) is moved between $h = 0$ and $h = 5$ m, and calculations may be partially clarified. We have considered reflective coefficient variations due to plants, snow, winter, or summer and scattering from auxiliary objects, which are nearby the positioner. Such objects include tower assemblies and different safe-guard obstructions. The utilization of diffraction fences is briefly explained. Numerical simulations have been successfully used both for design and optimizing of the rebuilt antenna range.

The observation of refractive index distributions for distance of 49.8 km, for various heights during the same day, reveals that distributions are very changeable. Therefore, we cannot speak correctly about stratum formulation. Nevertheless, we select for individual "layers" of PO simulations the same R_e effective radiuses, which were used for PEM simulations. Therefore the small parts of graphs are only displayed for comparison with measurements and PEM simulations.

It has been validated that the upgraded PO method offers more trustworthy calculations for low elevation propagations and diffraction zones. In this method, there is no supporting technique for special tropospheric situations for data transmission and communications together with electromagnetic compatibility. The small discrepancies could be incompletely clarified since permittivity, conductivity, and standard deviation change. Obviously, the selection of suitable effective radiuses, R_e, for individual heights could substantially diminish these discrepancies that could be very useful both for syntheses and analyses of various propagation phenomena including refraction and reflection from uneven ground. Obviously, this method provides more thoughtful results, when beams are narrow in horizontal plane such as pencil- or fan-shaped, which are used in radars.

The described PO method provides reliable computations for low-height fields and diffraction zones for numerous uneven terrains and realistic refractive index spreading. The used code for PO approximations allows only the utilization of one constant effective earth radius, R_e, for variable altitudes. This cannot be used for greater heights. Effects of the air refraction index, studied in part 5, could be neglected for coarse numerical simulations. However, the calculations could be more accurate using the recommendations ITU-R for higher altitudes of large-scale refractive effects.

Frequently, refractive propagation effects on electromagnetic wave propagation could be neglected especially for lower elevations. It is clearly demonstrated that we should consider ground scattering and we cannot only analyze the radiation pattern, refraction, and tropospheric losses.

Acknowledgements

The authors thank deceased Dr. D. Kupcak for very beneficial ideas during longtime collaboration both from theoretical and practical viewpoints.

The work was supported from ERDF/ESF "Cooperation in Applied Research between the University of Pardubice and companies, in the Field of Positioning, Detection and Simulation Technology for Transport Systems (PosiTrans)" (No. CZ.02.1.01/0.0/0.0/17_049/0008394).

Author details

Vladimir Schejbal*, Ondrej Fiser and Vadim Zavodny
University of Pardubice, Pardubice, Czech Republic

*Address all correspondence to: vladimir.schejbal@upce.cz

References

[1] Propagation by Diffraction. Recommendation ITU-R P.526-12. Geneva: 02/2012; p. 42

[2] Electrical characteristics of the surface of the earth. Recommendation ITU-R P.527-3. 2000. p. 5

[3] Kerr DE, editor. Propagation of Short Radio Waves. New York: McGraw-Hill; 1951. 763 p

[4] Beckmann P, Spizzichino A. The Scattering of Electromagnetic Waves from Rough Surfaces. Oxford: Pergamon Press; 1963. 503 p

[5] Meeks ML. Radar Propagation at Low Altitudes. Dedham: Artech; 1982. 105 p

[6] Allnutt JE. Satellite-to-Ground Radiowave Propagation. 2nd ed. London: The Institution of Engineering and Technology; 2011. 717 p

[7] Barclay L, editor. Propagation of Radiowaves. 3rd ed. London: The Institution of Engineering and Technology; 2013. 482 p

[8] Luebbers RJ. Propagation prediction for hilly terrain using GTD wedge diffraction. IEEE Transactions on Antennas and Propagation. 1984;**32**: 951-955. DOI: 10.1109/ TAP.1984.1143449

[9] Hviid JT, Anderson JB, Toftgård J, Bøjer J. Terrain-based propagation model for rural area-an integral equation approach. IEEE Transactions on Antennas and Propagation. 1995;**43**: 41-46. DOI: 10.1109/8.366349

[10] Akorli FK, Costa E. An efficient solution of an integral equation applicable to simulation of propagation along irregular terrain. IEEE Transactions on Antennas and Propagation. 2001;**49**:1033-1036. DOI: 10.1109/8.933482

[11] Cocheril Y, Vauzelle R. A new ray-tracing based wave propagation model including rough surfaces scattering. Progress in Electromagnetics Research. 2007;**75**:357-381. DOI: 10.2528/ PIER07061202

[12] Kuttler JR, Dockery GD. Theoretical description of the parabolic approximation/Fourier split-step method of representing electromagnetic propagation in the troposphere. Radio Science. 1991;**26**:381-393

[13] Papkelis EG et al. A radio-coverage prediction model in wireless communication systems based on physical optics and the physical theory of diffraction. IEEE Antennas and Propagation Magazine. 2007;**49**:157-165. DOI: 10.1109/MAP.2007.376622

[14] Schejbal V. Computing the electrical field strength of an antenna above an uneven earth (in Czech, Eng. Abs.). Slaboproudy Obzor. 1973;**34**:541-547

[15] Kupcak D. ATC Radar Antennas. Environment Influence on ATC Radar Operation (in Czech), vol. III. Prague: MNO; 1986, 358 p

[16] Schejbal V. Propagation over irregular terrain. Radioengineering. 1997;**6**:19-22. DOI: 10.13164/re.1997

[17] Schejbal V. Comparison of propagation over irregular terrain. Radioengineering. 1997;**6**:6-9. DOI: 10.13164/re.1997

[18] Schejbal V. Improved analysis of propagation over irregular terrain. Radioengieering. 2009;**18**:18-22. DOI: 10.13164/re.2009

[19] Schejbal V, Fiser O. Enhanced analysis of propagation over irregular terrain. In: Proceedings of the Fourth European Conference on Antennas and

Propagation (EuCAP); 12-16 April 2010; Barcelona, New York: IEEE. 2010. pp. 1697-1701

[20] Kupcak D. Evaluation of antenna radiation patterns by the generalized trapezoidal method. Tesla Electronics. 1974;**74**:43-50

[21] Schejbal V, Bezousek P. Computation accuracy of propagation over irregular terrain. In: Proceedings of the COMITE; Pardubice. 2001. pp. 95-98

[22] Schejbal V. Simplified computation of the electrical field strength of an antenna above an uneven earth. Slaboproudy Obzor. 1980;**41**:545-547 (in Czech, Eng. Abs.)

[23] Ufimtsev PY. New insight into the classical Macdonald physical optics approximation. IEEE Antennas and Propagation Magazine. 2008;**50**:11-20. DOI: 10.1109/MAP.2008.4563560

[24] Ufimtsev PY. Fundamentals of the Physical Theory of Diffraction. Hoboken: Wiley; 2007. p. 329

[25] Ufimtsev PY. Improved physical theory of diffraction: Removal of the grazing singularity. IEEE Transactions on Antennas and Propagation. 2006;**54**: 2698-2702. DOI: 10.1109/TAP.2006.882179

[26] Silver S. Microwave Antenna Theory and Design. New York: McGraw-Hill; 1949

[27] Skolnik MI. Radar Handbook. 3rd ed. New York: McGraw-Hill; 2008

[28] Skolnik MI. Introduction to Radar Systems. 2nd ed. New York: McGraw-Hill; 1980

[29] Hitney HV. Hybrid ray optics and parabolic equation methods for radar propagation modeling. In: Proceedings of the International Conference Radar; Brighton. 1992. pp. 58-61

[30] Nazari ME, Ghorbani A. Predicting a three-dimensional radar coverage area: Introducing a new method based on propagation of radio waves. IEEE Antennas and Propagation Magazine. 2016;**58**:28-34. DOI: 10.1109/MAP.2015.2501238

[31] Schejbal V et al. Czech radar technology. IEEE Transactions on Aerospace and Electronic Systems. 1994;**30**:2-17. DOI: 10.1109/7.250400

[32] Kupcak D. Electromagnetic field calculation over longitudinal terrain profile by partial wave method (in Czech). In: Proceedings of the Miteko; May 1980; Pardubice, Vol. 3. 1980. pp. 111-115

[33] Kupcak D. Surveillance radar location from terrain point of view (in Czech). In: Proceedings of the Miteko; May 1980; Pardubice, vol. 3. 1980. pp. 151-155

[34] The Radio Refractive Index: Its Formula and Refractivity Data. Recommendation ITU-R P.453-11. 07/2015. 25 p

[35] Effects of Tropospheric Refraction on Radiowave Propagation. Recommendation ITU-R P.834-5. 2005. 12 p

[36] Schejbal V, Zavodny V. Tropospheric propagation above uneven ground. Radioengieering. 2017;**26**: 972-978. DOI: 10.13164/re.2017.0972

[37] Bezousek P, Schejbal V. Bistatic and multistatic radar systems. Radioengieering. 2008;**17**:53-59. DOI: 10.13164/re.2008

[38] Shejbal T, Plsek R, Hermanek A. Comparison of azimuth estimation in PCL and MLAT systems applied on measured data. In: Proceedings of the

18th International Radar Symposium (IRS); 28-30 June 2017; Prague. 2017. pp. 1-9

[39] Cerny O, Dolecek R, Kopecky P, Schejbal V, Zavodny V. Optimization of far-field antenna range. Radioengieering. 2015;**24**:892-897. DOI: 10.13164/re.2015.0892

[40] Hollis JS, Lyon TJ, Clayton L. Microwave Antenna Measurements. 3rd ed. Atlanta: Scientific-Atlanta; 1985. 604 p

[41] Balanis CA. Modern Antenna Handbook. Hoboken: Wiley; 2008. 1680 p

[42] Schejbal V, Dolecek R, Zavodny V, Kopecky P, Fiser O. Far-field antenna range. In: Proceedings of the 14th Conference on Microwave Techniques (COMITE); 22-23 April 2015; Pardubice: IEEE. 2015. pp. 90-93

[43] Valtr P, Pechac P, Kvicera V, Grabner M. Estimation of the refractivity structure of the lower troposphere from measurements on a terrestrial multiple-receiver radio link. IEEE Transactions on Antennas and Propagation. 2011;**59**:1707-1715. DOI: 10.1109/TAP.2011.2122234

[44] Schejbal V, Grabner M. Investigation of propagation over terrain. In: Proceedings of the 24th International Conference Radioelektronika; 15-16 April 2014; Bratislava: IEEE. 2014. pp. 1-4

[45] Schejbal V, Dolecek R, Fiser O, Brazda V, Nemec Z. Propagation over terrain considering refractivity profiles. In: Proceedings of the 26th International Conference Radioelektronika; 19-20 April 2016; Košice: IEEE. 2016. pp. 481-485

[46] Schejbal V. Refraction effects for propagation over terrain. In: Proceedings of the 16th International Conference on Mathematical Methods in Electromagnetic Theory (MMET*2016); 5-7 July 2016; Lviv: IEEE. 2016. pp. 81-84

Chapter 8

Mode Interferences of VLF Waves in the Presence of an Anisotropic Terrestrial Waveguide

Ting Ting Gu and Hong Lei Xu

Abstract

It is known that the very low frequency (VLF) signals propagating at long distances over a terrestrial waveguide will experience a diurnal interference across the sunrise and sunset transitions. Early studies indicated that cascaded terrestrial waveguides can be applied to investigate the cycle-clip behaviors of amplitude and phase for VLF waves, with an abrupt effective height change by 20 km to represent the night and day transitions. Based on the knowledge, this chapter starts with some basic concepts on VLF wave propagation in the presence of an anisotropic terrestrial waveguide. Then, by performing the method of field matching at the discontinuous junctions, the coefficients of transmitted and scattered vectors are resulted from two derived matrix equations. Finally, the synthetic fields are calculated to simulate the VLF waves, which are subjected to an oblique geomagnetic field during long-distance propagation. The purpose of the present chapter is to illustrate the disturbing nature during sunrise and sunset for the VLF waves with periodic variations in amplitudes interfered by multimodes over long propagation paths.

Keywords: VLF waves, mode-conversion coefficient, sunrise transition, sunset transition, terrestrial cascaded waveguides

1. Introduction

The very low frequency (VLF) waves, namely, the radio waves whose wavelengths are between 10 and 100 km and operate at the frequencies ranging from 3 to 30 kHz, are usually used for long-distance propagations. In this frequency band, both the ground medium (occupied by land, lake, or seawater) and the ionosphere (the sky region of the height over 60–90 km distinct from daytime to nighttime states) are strongly reflecting for the excited waves, so that the VLF waves are conducted to propagate in free space between the two parallel reflecting walls. This is the so-called terrestrial waveguide. Generally, a vertical electric dipole antenna, polarized in perpendicular to the ground interface, can be idealized to generate these guiding waves in transverse magnetic (TM) modes, whereas the transverse electric (TE) modes generated by a horizontally polarized electric dipole antenna would not propagate along the radial direction. Assuming that the spherical coordinate system has been established, "the ground-ionosphere waveguide" is formed by the two concentric spherical shells.

Along a great propagation path across two states of the daytime and nighttime sections, the resulted cycle clips by interferences would be observed in phase for VLF radio waves [1–7]. Most of the early explanations of this phenomenon [8–10] were based on the ray-optic model; however, it seems unlikely to be accountable correctly for the mentioned matter for a long-distance propagation. That is because, on the one hand, it would be required to include lots of ray paths in explaining the VLF wave propagation at great distances, yet only a few low-order modes are necessary; and on the other hand, there should be consideration of including the diffraction corrections for using ray-optic models over a curved earth.

For a better knowledge of diurnal behaviors as phase and amplitude variations for VLF radio waves, extensive studies had been developed over the years by researchers. In the 1960s, the effect of multimode interferences had been addressed by Wait [11, 12] for VLF radio waves in the presence of a nonuniform earth-ionosphere waveguide. Later, the study had been extended by Bahar in the 1970s [13–16] where the inhomogeneous waveguide was assumed to be represented by successive cascaded earth-ionosphere waveguides, for which the height variations were accounted for the day-to-night transition. In these analyses, the mode-conversion coefficients had been formulated by employing the orthogonal properties of basis functions. Consequently, the resulted coefficients could be utilized to derive a set of coupled first-order differential equations for the forward and backward wave amplitudes in the transition region, to extend the study into a more general case with arbitrarily varying effective height. However, it should be pointed out that these computational schemes by previous treatments had neglected the influence of the geomagnetic field. In the study by Galejs [17, 18], the transition geometry of waveguide with an abrupt height change of dh was equivalent to that of a waveguide with identical height boundaries but different surface impedances, representing the anisotropic ionosphere in daytime and nighttime, respectively. However, the author also indicated the derived reflection coefficients would be of even smaller significance in a more realistic problem of gradual height or impedance changes. Therefore, the mode-conversion coefficients remain to be formulated for higher accuracy of the VLF radio waves to include the influence of geomagnetic field at great propagation path across daytime and nighttime states.

The analytical formulas for mode-conversion coefficients have been derived in the circadian transitional period by Pan et al. [19–21] applied in ELF/SLF electromagnetic wave propagation in earth-ionosphere waveguide due to sunrise and sunset. Apart from its simplicity, the proposed computations are also feasible for studying the influence of the anisotropic behavior from the ionosphere, including reflection characteristics of the transitional daytime and nighttime sections. The study leads us to treat the ionosphere as anisotropic plasma medium and derive the mode-conversion coefficients resulted from the daytime and nighttime transitions for VLF waves.

In this chapter, the matter of VLF wave propagation is addressed in the presence of an earth-ionosphere cavity. Considering the diurnal behavior caused by sunrise and sunset transitions, the mode-conversion coefficients in an earth-ionosphere waveguide have been derived under an abrupt effective height change. The proposed computational scheme guarantees satisfactory accuracy of the electromagnetic field in the transition ranges and can be developed in the derivation of the coupled differential equations for the wave amplitudes in a transition section of gradually varying height. Additionally, the mode interference has been analyzed. The time dependence $e^{i\omega t}$ is suppressed throughout the analysis.

2. VLF radio wave propagation in the presence of an anisotropic earth-ionosphere waveguide

In this section, the expressions of electromagnetic components have been addressed for VLF radio waves propagating in an anisotropic earth-ionosphere waveguide. The computations are based on an analysis of exploiting the concept that the VLF radio waves propagate in a spherical earth-ionosphere waveguide, with finite surface impedance boundaries. Wave components are formulated in the expression of Airy functions for the VLF radio waves in the context.

2.1 Formulation

Assume that the strength of geomagnetic field B_0 is suppressed by an oblique incident angle θ_0 in $\hat{z}$ direction (the direction perpendicular to the interface of the earth) in the $\hat{x}$-$\hat{z}$ plane. The ionosphere above the air is regarded as the homogeneous anisotropic plasma, characterized by a tensor permittivity $\acute{\varepsilon}$ [22], which is defined by

$$\acute{\varepsilon}_{A,B} = \varepsilon_0(I + M_{A,B}) \tag{1}$$

in which the constant ε_0 represents for the free-space permittivity, I is a 3×3 unit matrix, and the susceptibility of ionosphere M can be expressed as follows:

$$M = \frac{-X}{U(U^2 - y^2)} \times \begin{bmatrix} U^2 - l^2y^2 & inUy & -lny^2 \\ -inUy & U^2 & ilUy \\ -lny^2 & -ilUy & U^2 - n^2y^2 \end{bmatrix} \tag{2}$$

where the variable U is the effective electron collision frequency of the ionosphere, defined by $U = 1 + i\nu/\omega$, $y = \omega_H$, and $X = \omega_0^2/\omega^2$; ω_H and ω_0 are the gyrofrequency of the electrons and angular plasma frequency of the ionosphere, respectively; and l and n are the directional cosines of the geomagnetic field in the $\hat{x}$ and $\hat{z}$ directions, respectively.

2.2 Expressions of the components for VLF radio waves in an anisotropic terrestrial waveguide

In what follows, it is assumed that the characteristics of VLF radio waves are determined by propagation paths from the exciting source to the receiving field point, that is, the characteristics of propagation will not be affected by other propagation paths.

2.2.1 Computational expressions

In an idealized geometry, the ionosphere layers are regarded as the anisotropic plasma, and the waveguide is occupied by free space, characterized by the permeability μ_0 and uniform permittivity ε_0. The spherical earth is characterized by the permeability μ_0, uniform permittivity ε_g, and conductivity σ_g. Considering the anisotropic properties of the ionosphere, the wave components for VLF radio waves in the earth-ionosphere waveguide would be no longer strictly separated by TM waves and TE waves [23, 24]. Each wave type will consist of six electromagnetic wave components expressed in the following form:

$$\begin{matrix} E_{rn} \\ E_{\theta n} \\ E_{\phi n} \\ H_{rn} \\ H_{\theta n} \\ H_{\phi n} \end{matrix} = a_n \left\{ \begin{matrix} F_n(t_n, y) \\ i\beta_n^{-1} F'_n(t_n, y)/k \\ \Delta_{Emn} g_n(t_n, y)\beta_n^{-1} \\ \Delta_{Emn} g_n(t_n, y) \\ i\Delta_{Emn} g'_n(t_n, y)\beta_n^{-1}/k \\ -\beta_n^{-1} F_n(t_n, y) \end{matrix} \right\} e^{ika\beta_n\theta} \quad (3)$$

in which the functions $F_n(z)$ and $g_n(z)$ stand for the normalized "height-gain" functions for the TE mode and TM mode, respectively, defined by

$$F_n(y) = \frac{w_2(t_n - y) + B_1(t_n)w_1(t_n - y)}{w_2(t_n) + B_1(t_n)w_1(t_n)} \quad (4)$$

$$g_n(y) = \frac{w_2(t_n - y) + B_2(t_n)w_1(t_n - y)}{w_2(t_n) + B_1(t_n)w_1(t_n)} \quad (5)$$

in which

$$B_1(t_n) = \frac{-w'_2(t_n) - qw_2(t_n)}{w'_1(t_n) - qw_1(t_n)}; \quad B_2(t_n) = \frac{-w'_2(t_n) - q^h w_2(t_n)}{w'_1(t_n) - q^h w_1(t_n)} \quad (6)$$

with

$$q = i\left(\frac{ka}{2}\right)^{\frac{1}{3}} \Delta_g; q^h = i\left(\frac{ka}{2}\right)^{\frac{1}{3}} \frac{1}{\Delta_g} \quad (7)$$

In the above formulas, the variable β_n is the nth wave impedance defined by $\beta_n = 1 + t_n[2/(ka)]^{2/3}/2$, and the sequential roots t_n are determined by the modal equation in [5], rewritten as follows:

$$\begin{aligned} &\left[w'_2(t_n - y_0) + i\Delta_{11}\left(\frac{ka}{2}\right)^{\frac{1}{3}} w_2(t_n - y_0)\right. \\ &\left. + B_1(t_n)\left(w'_1(t_n - y_0) + i\Delta_{11}\left(\frac{ka}{2}\right)^{\frac{1}{3}} w_1(t_n - y_0)\right)\right] \\ &\times\left\{\Delta_{22}w'_2(t_n - y_0) - i\left(\tfrac{ka}{2}\right)^{\frac{1}{3}} w_2(t_n - y_0)\right. \\ &\left. + B_2(t_n)\left[\Delta_{22}w'_1(t_n - y_0) - i\left(\tfrac{ka}{2}\right)^{\frac{1}{3}} w_1(t_n - y_0)\right]\right\} \\ &- i\left(\tfrac{ka}{2}\right)^{\frac{1}{3}} \Delta_{12}\Delta_{21}\left[w'_2(t_n - y_0) + B_2(t_n)w'_1(t_n - y_0)\right] \\ &\times\left[w_2(t_n - y_0) + B_1(t_n)w_1(t_n - y_0)\right] = 0 \end{aligned} \quad (8)$$

The coupling impedance Δ_{Emn} is expressed as

$$\Delta_{Emn} = M_n$$
$$\left\{ i\Delta_{21}\left(\frac{ka}{2}\right)^{\frac{1}{3}}\left[w_2(t_n - y_0) + B_1(t_n)w_1(t_n - y_0)\right]\right\}$$
$$\times\left\{\Delta_{22}w_2'(t_n - y_0) - i\left(\tfrac{ka}{2}\right)^{\frac{1}{3}}w_2(t_n - y_0)\right.$$
$$\left.+B_2(t_n)\left[\Delta_{22}w_1'(t_n - y_0) - i\left(\tfrac{ka}{2}\right)^{\frac{1}{3}}w_1(t_n - y_0)\right]\right\}^{-1} \tag{9}$$

and the nth-mode wave admittance u is

$$Y_n = \beta_n/kZ_0 \tag{10}$$

where the variable Z_0 stands for the free-space wave impedance.

2.2.2 Definition of the airy functions

In the evaluation of Eq. (8) to Eq. (9), the Airy functions have been employed. In order to clarify these special functions, the following expressions provided by the use of Bessel functions are defined as follows [23]:

$$w_1(x) = e^{-i\frac{2\pi}{3}}\left(\frac{-\pi t}{3}\right)^{\frac{1}{2}}H_{1/3}^{(2)}\left[\frac{2}{3}(-t)^{\frac{3}{2}}\right] \tag{11}$$

$$w_2(x) = e^{+i\frac{2\pi}{3}}\left(\frac{-\pi t}{3}\right)^{\frac{1}{2}}H_{1/3}^{(1)}\left[\frac{2}{3}(-t)^{\frac{3}{2}}\right] \tag{12}$$

$$w_1'(x) = e^{-i\frac{\pi}{3}}\left(\frac{\pi}{3}\right)^{\frac{1}{2}}(-t)H_{2/3}^{(2)}\left[\frac{2}{3}(-t)^{\frac{3}{2}}\right] \tag{13}$$

$$w_2'(x) = e^{+i\frac{\pi}{3}}\left(\frac{\pi}{3}\right)^{\frac{1}{2}}(-t)H_{2/3}^{(1)}\left[\frac{2}{3}(-t)^{\frac{3}{2}}\right] \tag{14}$$

in which the Airy functions are defined by the first- and second-kind Hankel functions of one third or two third order. In addition, the Wronskian equality for the Airy function is defined as follows:

$$W = w_1w_2' - w_2w_1' = constant. \tag{15}$$

2.3 Computation and discussion

In what follows, the propagation parameters are computed for VLF waves and quantitatively compared to the isotropic case to study the anisotropic ionosphere properties influenced by the geomagnetic fields.

2.3.1 Impedance matrix of ionosphere

To consider an anisotropic ground-ionosphere waveguide, the surface impedance matrix for the field components has been applied to terminate the source-free

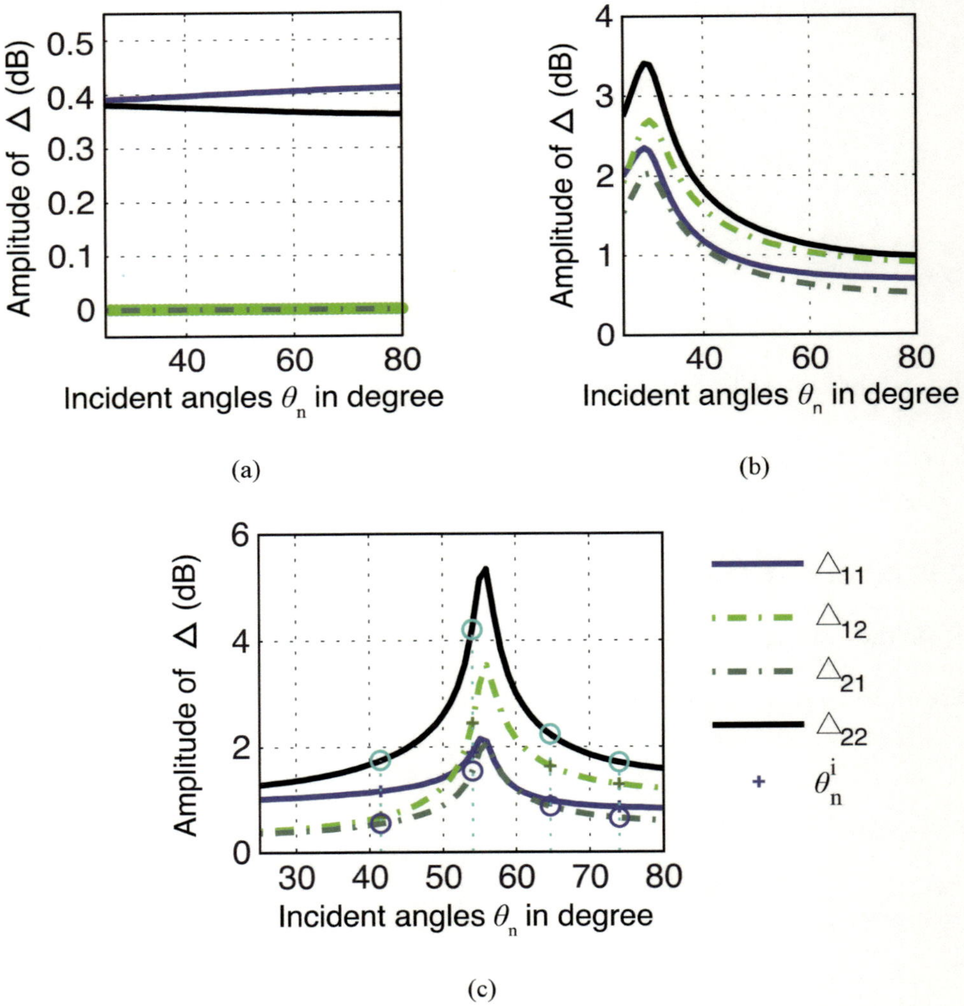

Figure 1.
The amplitudes of elements of the ionospheric impedance matrix versus the propagation direction $\boldsymbol{\theta_n}$*, at* f = *10 kHz, with strength of earth geomagnetic field: (a)* **10**$^{-6}$ G*, (b)* **0.45** G*, (c)* **0.5** G*, respectively.*

waveguide at the same altitude as in the region of the waveguide source. In **Figures 1** and **2**, the elements of impedance matrix have been computed at an operating frequency by $\boldsymbol{f} = \mathbf{10}\boldsymbol{kHz}$ for anisotropic ionosphere, where the equivalent effective height of the ionosphere is assumed to be $\boldsymbol{h} = \mathbf{90}\boldsymbol{km}$ for summer daytime, electron density is N = $\mathbf{10^9}$ m^{-3}, and electron collision frequency is ν = $\mathbf{10^7}$ s^{-1}. And the ground (assumed to be occupied by seawater) is characterized by the relative dielectric constant $\boldsymbol{\varepsilon_r} = \mathbf{80}$ and the conductivity $\boldsymbol{\sigma} = \mathbf{5}$ Sm^{-1}. In the numerical examples, the incident angles $\boldsymbol{\theta_n}$ are selected, ranging from 10 to 80°, while the strength of earth geomagnetic field $\boldsymbol{B_0}$ is chosen by $\mathbf{10^{-6}}$ G, 0.45 G, and 0.5 G, respectively, where the inclination angle is characterized by a dip angle with $\boldsymbol{\vartheta} = \mathbf{0}$, in east-to-south direction.

The diagonal elements $\boldsymbol{\Delta_{11}}$ and $\boldsymbol{\Delta_{22}}$ in the matrix of ionosphere surface impedance are accounted for the TM mode and TE mode, respectively, and the elements of ionosphere impedance matrix are affected by the earth geomagnetic field. Specifically, the curves of TM mode and TE mode are interfered more as suppressed by stronger earth geomagnetic field strengths.

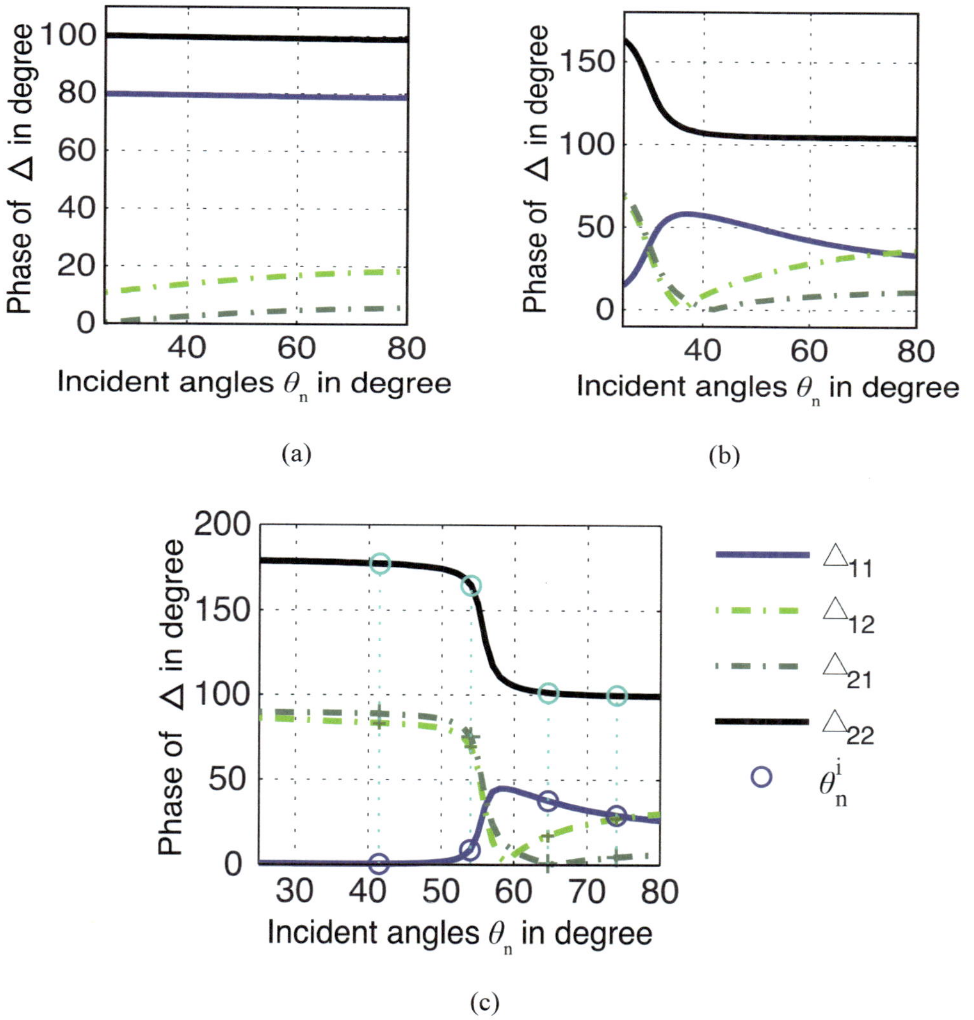

Figure 2.
The phases of elements of the ionospheric impedance matrix versus the propagation direction θ_n, at f = 10 kHz, with strength of earth geomagnetic field: (a) 10^{-6} G, (b) 0.45 G, (c) 0.5 G, respectively.

2.3.2 Attenuations and velocities of the VLF radio waves

At VLF radio frequencies, the surface impedance of the ground is much smaller than that of the ionosphere. This makes the change of the ground conductivity to have little effect on the phase velocity and attenuation rate of the waves [19]. In the following computations, the ground is assumed to be idealized to the sea surface. By the proposed formulas, we will compute the anisotropic ionosphere parameters. To begin with, the modal equation of VLF wave propagation can be solved readily by (8), so that the characteristic parameters are calculated correspondingly. In **Figures 3(a)** and **3(b)**, with the conductivity and the relative dielectric constant of sea water being $\sigma_g = 4$ S/m and $\varepsilon_{rg} = 80$, respectively, the attenuation rates α (dB/1000 km) are calculated with respect to the normalized frequencies (with $f_c = 10kHz$) where the boundary of the lower ionosphere is assumed at an altitude $h = 70$ km and $h = 90$ km for daytime and nighttime, respectively.

As is depicted from **Figure 3(a)**, the attenuation rates of lower modes are raised by increasing the operating frequencies. But for the higher modes, the attenuation

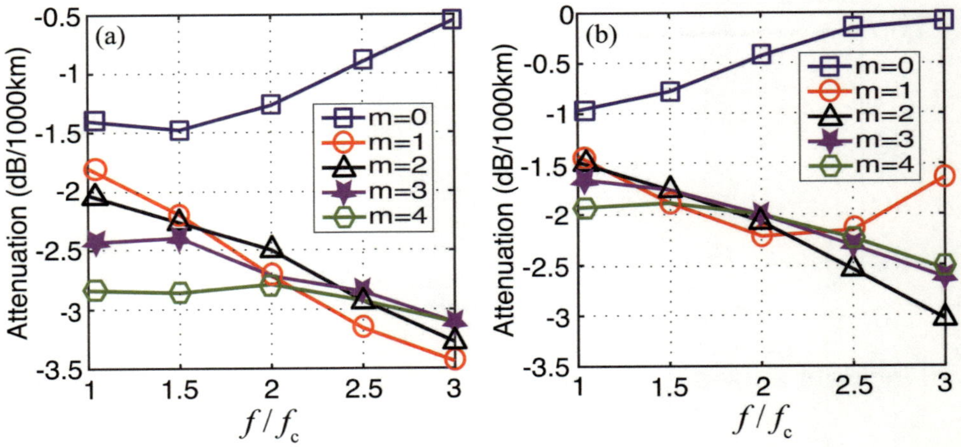

Figure 3.
The attenuation rates versus the normalized frequency f/f_c with (a) $h = 70km$ and (b) $h = 90km$, respectively.

rates decrease on the same condition. This is because the fundamental mode as the ground wave term cannot propagate at great propagation distances, whereas the high modes as sky waves, multi-reflected by the ground and ionosphere walls, will propagate through long path. Considering that both the radiating antenna and the receiving antenna are located in the same state of daytime or nighttime, the equivalent altitude of the ionosphere is quite different from each state along with propagation path. In **Figure 3**, it is shown that the attenuation rates of lower modes have different behaviors for daytime and nighttime [$m = 0$ in **Figure 3(a)** for daytime and $m = 0, 1$ in **Figure 3(b)** for nighttime], respectively, than the other modes as increased by operating frequency. Generally, it is seen that the higher effective ionosphere height is, the lower modes appear with attenuation rates as increased by frequencies.

As is shown in **Figures 4(a)** and **4(b)**, the relative phase velocities are evaluated, correspondingly, in variance of the normalized frequencies (with $f_c = 10kHz$) for daytime and nighttime, suppressed by earth geomagnetic field with the strength chosen as 0.45 and 0.5 Gauss, respectively. The formulas for VLF wave components in the presence of earth-ionosphere waveguide have been given in [12], as well as their special dispersive features by the propagation coefficients. It is noted that the relative velocities of VLF radio waves decrease as increased by operating frequencies. However, it is also known from **Figures 4(a)** and **(b)** that the phase velocity of the fundamental mode, standing for the ground wave, is affected very little by earth geomagnetic inclination angles, while that of the high modes changes greatly by the angles. Specifically, the influence of the relative phase velocity resulted from the earth geomagnetic field is about in the range of 1–3% [19].

In conclusion, the attenuation rates and phase velocities of the ground wave mode are affected by strength and inclination angles of earth geomagnetic field. However, at a greater geomagnetic angle, the attenuation rate becomes larger, and the relative phase velocity decreases correspondingly. On the other hand, the influences of the attenuation rates and the phase velocities are weakened by propagating direction. For fundamental modes, there exist directional actions in attenuation rate and phase velocity. For example, the attenuation rate propagating westward will be greater than that propagating eastward, if the relative phase velocity propagating eastward is also smaller than that propagating westward [19].

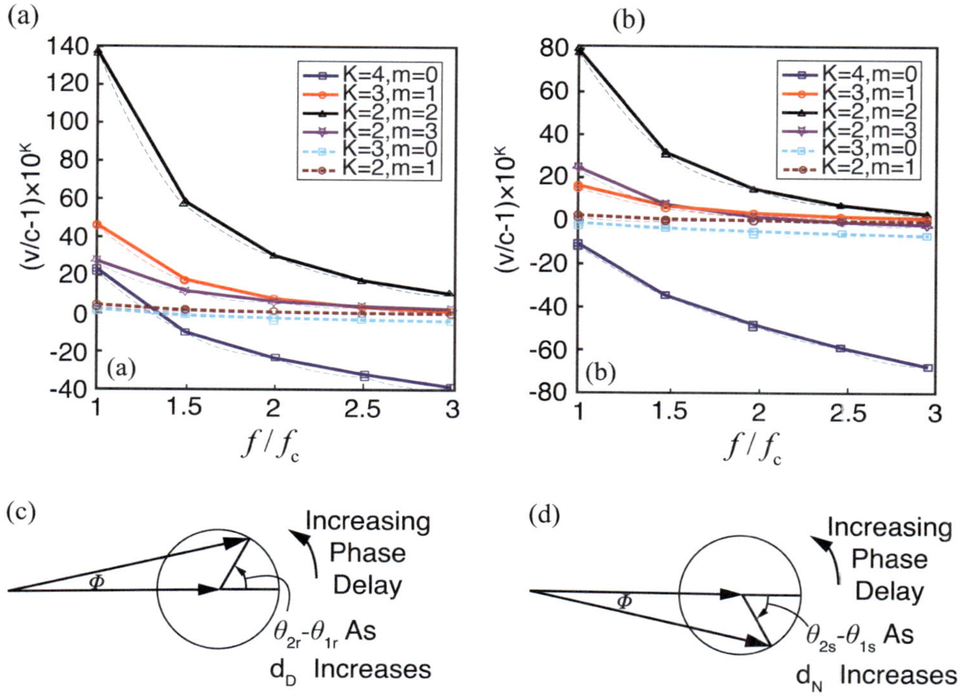

Figure 4.
The relative phase velocity versus the relative frequency f/f_c with (a) $h = 70km$, (b) $h = 90km$, (c) phase delay of d_{D-N}, and (d) phase delay of d_{N-D}. Solid lines (——) are for earth geomagnetic inclination angle of 80°, and dashed lines (----) are for earth geomagnetic field inclination angle of 90°.

3. Mode interferences of VLF waves in an anisotropic waveguide due to sunrise and sunset

In this section, the orthogonal properties of basis functions are considered at each side of daytime and nighttime waveguides, respectively, while the ionosphere layer has been treated as anisotropic plasma. The transmission and reflection coefficients are determined by the resulting linear equations from field matching. Thus, the mode-conversion coefficients are obtained for VLF waves propagating in the terrestrial waveguide during sunrise and sunset.

3.1 Mode-conversion coefficients across abrupt discontinuity

The idealized model representing the sunrise or sunset transition is depicted in **Figure 5**, where a cascading anisotropic waveguide is formed by two sections with different effective heights. Each section of the ionosphere layer is regarded as anisotropic plasma, and the waveguide is occupied by the free space, characterized by the permeability μ_0 and the uniform permittivity ε_0. The spherical earth is characterized by μ_0, uniform permittivity of ε_g, and conductivity of σ_g. In **Figure 5**, the effective height of ionosphere $h(x)$ is given by a unit step function, defined by $h(x) = h_A + (h_B - h_A)u(x)$. Thus, at the junction with $x' = 0$, it can be analyzed as a two-port waveguide. In region A, the electromagnetic field can be expressed by the summation of excited wave modes. Let a^A and b^A be vectors in terms of the amplitudes of forward wave and backward wave vectors, respectively, and $\tilde{F}^A$ and $\tilde{G}^A$ be row vectors representing the height-gain functions of TM mode and TE mode, respectively.

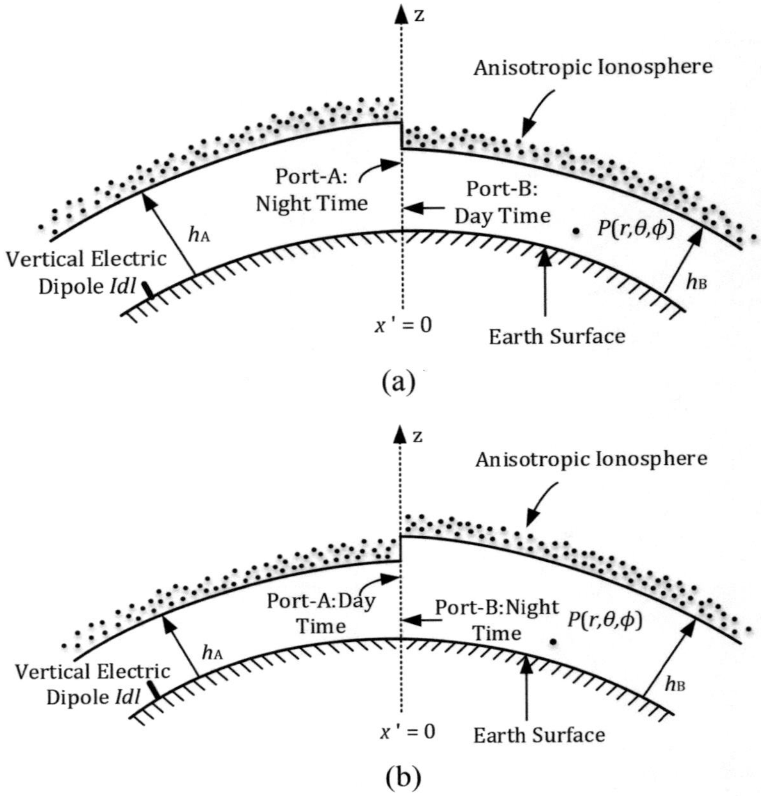

Figure 5.
Geometry of transitional regions in the nonuniform terrestrial waveguides with an abrupt effective height change of 20 km caused by sunset and sunrise, respectively. (a) Night-to-day transition; (b) day-to-night transition.

In region $A(x' > 0)$, the excited wave modes are expressed as follows:

$$\left.\begin{aligned} E_r &= \sum a_n^A F_n^{+A}(z) e^{ik\beta_n^{+A} x'} + \sum b_n^A F_n^{-A}(z) e^{-ik\beta_n^{-A} x'} \\ E_\phi &= \sum a_n^A G_n^{+A}(z) \left[\beta_n^{+A}\right]^{-1} e^{ik\beta_n^{+A} x'} \\ &\quad - \sum b_n^A G_n^{-A}(z) \left[\beta_n^{-A}\right]^{-1} e^{-ik\beta_n^{-A} x'} \\ H_r &= \sum a_n^A G_n^{+A}(z) e^{ik\beta_n^{+A} x'} + \sum b_n^A G_n^{-A}(z) e^{-ik\beta_n^{-A} x'} \\ H_\phi &= -\sum a_n^A F_n^{+A}(z) \left[\beta_n^{+A}\right]^{-1} e^{ik\beta_n^{+A} x'} \\ &\quad + \sum b_n^A F_n^{-A}(z) \left[\beta_n^{-A}\right]^{-1} e^{-ik\beta_n^{-A} x'} \end{aligned}\right\} \tag{16}$$

In Eq. (16), the different phase velocities, attenuation rates, and height functions are taken into consideration for the incident wave and the reflected wave for each wave type, respectively. In region $B(x' > 0)$, we have

$$\left.\begin{aligned} E_r &= \sum a_n^B F_n^{+B}(y) e^{ik\beta_n^{+B} x'} \\ E_\phi &= \sum a_n^B G_n^{+B}(y) \left[\beta_n^{+B}\right]^{-1} e^{ik\beta_n^{+B} x'} \\ H_r &= \sum a_n^B G_n^{+B}(y) e^{ik\beta_n^{+B} x'} \\ H_\phi &= -\sum a_n^B F_n^{+B}(y) \left[\beta_n^{+B}\right]^{-1} e^{ik\beta_n^{+B} x'} \end{aligned}\right\} \tag{17}$$

$$G_n(z) = \Delta_{Emn} g_n(z) \tag{18}$$

in which $F_n(z)$ and $G_n(z)$ are the elements of the column vector of height-gain functions.

The superscript B is designated instead of A by different matrix quantities. The element $Y_n(z)$ represents the main element of diagonal matrix for the wave admittance.

At the junctions of day and night sections, the interferences of VLF radio waves occur due to discontinuity of abrupt effective heights. Assume that the electromagnetic waves are incident at the junction from port A in **Figure 1(a)**:

$$a^B = S^{BA} a^A; b^A = S^{AA} a^A \tag{19}$$

The symbol S^{BA} is determined by a square transmission scattering matrix, whose element S_{nm}^{BA} is the complex amplitude of the nth transmitted mode through the junction at $x' = 0$ from port A. And the symbol S^{AA} is a square reflection scattering matrix, whose element S_{nm}^{AA} is determined by the complex amplitude of the nth reflected mode when the mth mode of unit amplitude is incident at the junction $x' = 0$ from port A. Similarly, the superscripts A and B are interchanged as the above quantities related to wave incidented from port B.

3.1.1 Field matching

At the junction of the waveguide sections with $(x' = 0)$, the boundary condition is applied as follows:

$$\tilde{F}^{+A} a^A + \tilde{F}^{-A} S^{AA} a^A = \tilde{F}^{+B} S^{BA} a^A \tag{20}$$

$$\tilde{G}^{+A} Y^{+A} a^A - \tilde{G}^{-A} Y^{-A} S^{AA} a^A = \tilde{G}^{+B} Y^{+B} S^{BA} a^A \tag{21}$$

$$\tilde{G}^{+A} a^A + \tilde{G}^{-A} S^{AA} a^A = \tilde{G}^{+B} S^{BA} a^A \tag{22}$$

$$\tilde{F}^{+A} Y^{+A} a^A - \tilde{F}^{-A} Y^{-A} S^{AA} a^A = \tilde{F}^{+B} Y^{+B} S^{BA} a^A \tag{23}$$

in which the column vectors Y^A and Y^B stand for the characteristic admittance diagonal matrices for the propagating modes in section A and section B, respectively. The superscript "+" or "−" represents the directed wave or reflected wave at each port, respectively.

3.1.2 VLF waves across abrupt junctions

Taking into account the properties of height-gain functions, the row vectors $\tilde{F}^A$ and $\tilde{G}^A$ at port A are orthogonal over the interval $(0 \le h \le h_A)$. Pre-multiply (20) and (22) by the column vectors F^{+B} and $\rho\tilde{G}^B$, and then integrate over the interval $(0, h_A)$, respectively; add up them. Similarly, pre-multiply (21) and (23) by the column vectors F^{+B} and $\rho\tilde{G}^B$, and then integrate over the interval $(0, h_A)$, respectively; add up them. It is noted that the vector a^A can be chosen arbitrarily and ρ is defined by $\rho = \Delta_{12}/\Delta_{21}$; it can be obtained that

$$\left(C^{++B\ \ A} + D^{++B\ \ A}\right) + \left(C^{+-B\ \ A} + D^{+-B\ \ A}\right)S^{AA} = W^B S^{BA} \tag{24}$$

$$W^A Y^{+A} - W^{+-A\ \ A} Y^{-A} S^{AA} = C^{++A\ \ B} Y^{+B} S^{BA} \tag{25}$$

in which matrix $W^{A,B}$ are diagonal matrices because of the orthogonal property of the height functions. Finally, we have

$$S^{BA} = \left[W^B\right]^{-1}\left[C^{++B\ \ A} + D^{++B\ \ A} + \left(C^{+-B\ \ A} + D^{+-B\ \ A}\right)S^{AA}\right] \tag{26}$$

$$\begin{aligned} S^{AA} = &\left\{W^{+-A\ \ A} Y^{-A} + C^{++A\ \ B} Y^{+B}\left[W^B\right]^{-1}\left[C^{+-B\ \ A} + D^{+-B\ \ A}\right]\right\}^{-1} \\ &\times\left\{W^A Y^{+A} - C^{++A\ \ B} Y^{+B}\left[W^B\right]^{-1}\left[C^{++B\ \ A} + D^{++B\ \ A}\right]\right\} \end{aligned} \tag{27}$$

It yields

$$E - X \approx [E + X]S^{AA} \tag{28}$$

in which

$$X \approx \left[Y^{-A}\right]^{-1}\left[W^{+-A\ \ A}\right]^{-1} C^{++B\ \ A} Y^{+B}\left[W^B\right]^{-1}\left(C^{+-B\ \ A} + D^{+-B\ \ A}\right) \tag{29}$$

Therefore, the reflection scattering matrix can be approximated by

$$S^{AA} = \Delta^{AA}\left(E - \Delta^{AA}\right)^{-1} = \sum_{k=0}^{\infty}\left(\Delta^{AA}\right)^k \tag{30}$$

with

$$\Delta^{AA} = \frac{1}{2}(E - X) \approx \frac{1}{2}\left(E - \left[Y^{-A}\right]^{-1}\left[W^{+-A\ \ A}\right]^{-1} C^{++B\ \ A} Y^{+B}\left[W^B\right]^{-1}\left(C^{+-B\ \ A} + D^{+-B\ \ A}\right)\right) \tag{31}$$

where the elements of the matrices $\overset{+\pm}{C^B_{nm}}$, $\overset{+\pm}{D^B_{nm}}$, and $W^{A,B}_{nn}$ are obtained by

$$\int_0^{h_A}\left[F_m^{+A}(y)F_n^{\pm B}(y)+\rho G_m^{+A}(y)G_n^{\pm B}(y)\right]dy$$

$$C_{nm}^{+\pm B} = \int_0^{h_A}\left[F_n^{+B}(y)F_m^{\pm A}(y)+\rho G_n^{+B}(y)G_m^{\pm A}(y)\right]dy \quad C_{mn}^{+\pm A} \tag{32}$$

and

$$\begin{aligned} D_{nm}^{+\pm B} &= \int_{h_A}^{h_B}\left[F_n^{+B}(y)F_m^{\pm A}(y)+\rho G_n^{+B}(y)G_m^{\pm A}(y)\right]dh \\ &\approx \left[F_n^{+B}(y)F_m^{\pm A}(y)+\rho G_n^{+B}(y)G_m^{\pm A}(y)\right]_{h_A}(h_B-h_A) \\ &\quad \frac{+(h_B-h_A)^2}{2}\left[F_n^{+B\,\prime}(y)F_m^{\pm A}(y)+\rho G_n^{+B}(y)'G_m^{\pm A}(y) \right. \\ &\quad \left. +F_n^{+B}(y)F_m^{\pm A\,\prime}(y)+\rho G_n^{+B}(y)G_m^{\pm A\,\prime}(y)\right]_{h_A} \end{aligned} \tag{33}$$

where

$$W_{nn}^{A} = \int_0^{h_A}\left[\left(F_n^{+A}(y)\right)^2+\rho^{+A}\left(G_n^{+A}(y)\right)^2\right]dy \tag{34}$$

$$W_{nn}^{B} = \int_0^{h_B}\left[\left(F_n^{+B}(y)\right)^2+\rho^{+B}\left(G_n^{+B}(y)\right)^2\right]dy \tag{35}$$

in which $F_n^{+B}(y)$ and $G_n^{+B}(y)$ are defined by Eq. (4) to Eq. (5), respectively. It is noted that for the case where the ionosphere may be considered as isotropic, the resulted coefficients can be reduced to the early analysis [13]. It is known from Eq. (23) that the reflection scattering matrix can be approximated in the form of

$$\Delta^{AA} \approx \sum_{p=0}^{\infty}\left(\frac{1}{2}(E-X)\right)^k \tag{36}$$

whose elements are

$$S_{nm}^{AA} \approx \Delta_{nm}^{AA} = \frac{1}{2}\left(\delta_{nm} - \sum_k \frac{\beta_k^B C_{np}^{++B\ \ A}\left(C_{pm}^{+-B\ \ A} + D_{pm}^{+-B\ \ A}\right)}{\beta_k^A W_{nn}^A W_{kk}^B}\right) \tag{37}$$

in which the delta function δ_{nm} is defined by an impulsive-like function with $\delta_{nn} = 1(n = 1)$. Correspondingly, the element of transmission scattering matrix can be obtained as follows:

$$S_{nm}^{BA} \approx \frac{C_{nm}^{++A \quad B}}{W_{nn}^{B}} + \sum_{k=0}^{\infty} \frac{\left(C_{nk}^{+-A \quad B} + D_{nk}^{+-A \quad B}\right)}{W_{nn}^{B}} S_{km}^{AA} \tag{38}$$

It is noted that Eq. (37) and Eq. (38) define the reflection matrix S^{AA} and scattering matrix S^{BA} are analogous to the definitions in the transmission line theory, suppressed by the normalized load admittance.

3.2 VLF waves across cascaded waveguides

Due to the revolution of the earth, the boundary of daytime and nighttime regions is moving over time. Thus, in the period of the daytime and nighttime transitions, the received phase of VLF radio waves in a receiving antenna appears a regular fluctuation. This is called the transition period interfered by mode conversions.

In **Figure 6**, the geometries of the transition model are presented by cascaded terrestrial waveguides, respectively. These models are approximated by an inhomogeneous waveguide represented in **Figure 7** with a varying height. The height function can be defined as follows:

$$h(x) = \frac{1}{2}(h_N + h_D) + \Delta h \cdot cos\left(\frac{\pi u}{L}\right) \tag{39}$$

in which

$$u(x) = \left(1 - 2\frac{\alpha}{\pi}\right)x + L \cdot \frac{\alpha}{\pi} + \left(2\frac{\alpha}{\pi} - 1\right)x_N \tag{40}$$

with $x = x_N + \Delta x$.

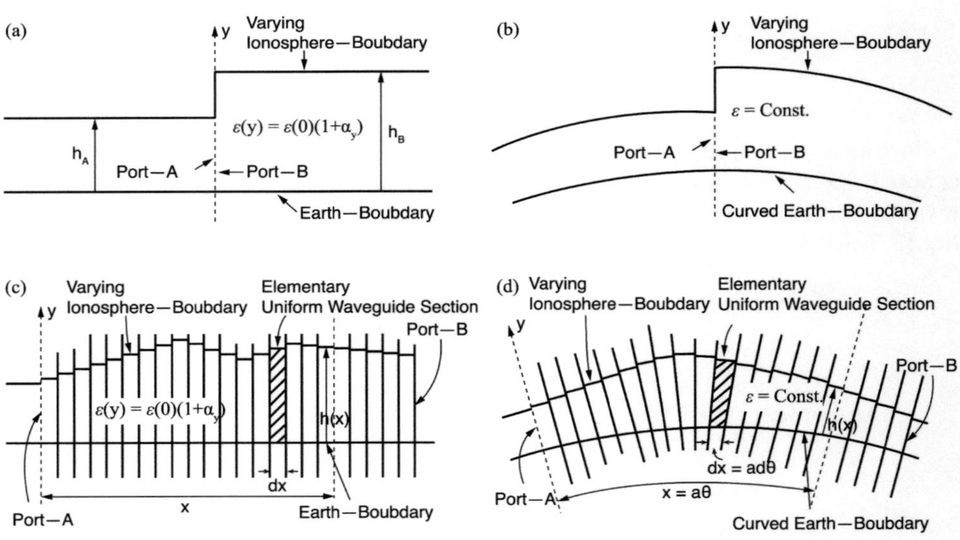

Figure 6.
Varying height profile in the transition region between the night and day paths with (a) planar model in abrupt discontinuity; (b) spherical model in discontinuity; (c) cascaded planar model in abrupt discontinuity; and (d) cascaded spherical model in discontinuity.

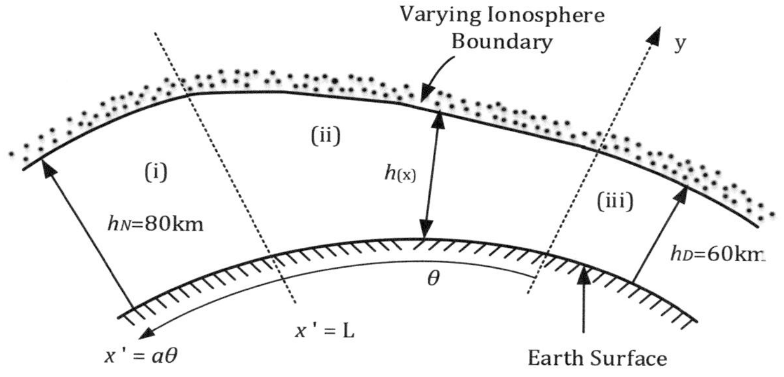

Figure 7.
Gradually varying height profile in the transition region between the night and day paths with (i) nighttime section, (ii) transition section with varying ionospheric effective height, and (iii) daytime section.

In this case, the scattered coefficients derived in Eq. (36) to Eq. (37) can be developed as follows:

$$S_{nm}^{BA} = \left[C_{nm}^{++B\quad A} + D_{nm}^{++B\quad A} + \left(C_{nm}^{+-B\quad A} + D_{nm}^{+-B\quad A}\right)S_{nm}^{AA}\right]/W_{nn}^{B} \tag{41}$$

So that, we have

$$\begin{aligned} \frac{dS_{nm}^{BA}}{dx} &= \frac{S_{nm}^{BA}}{dh}\frac{dh}{dx}\bigg|_{h_A=h_B=h(x)} \\ &- \left[\frac{dC_{nm}^{++B\quad A}}{dh} + \frac{dD_{nm}^{++B\quad A}}{dh} + \frac{d\Delta_{nm}^{AA}}{dh}\right]\frac{dh}{dx}\bigg|_{h_A=h_B=h(x)} \end{aligned} \tag{42}$$

and

$$\begin{aligned} \frac{S_{nm}^{AA}}{dx} &= \frac{S_{nm}^{AA}}{dh}\frac{dh}{dx}\bigg|_{h_A=h_B=h(x)} \\ &\frac{d\Delta_{nm}^{AA}}{dh}\frac{dh}{dx}\bigg|_{h_A=h_B=h(x)} \end{aligned} \tag{43}$$

in which

$$\begin{aligned} \frac{d\Delta_{nm}^{AA}}{dh}\Big|_{h(x)}\frac{dh}{dx} &= \frac{S_{nm}^{AA}}{dx} \\ &- \left[\frac{dC^{++B\quad A}}{dh} + \frac{1}{2}\frac{dD^{++B\quad A}}{dh}\right]\frac{dh}{dx}\bigg|_{h_A=h_B=h(x)} \end{aligned} \tag{44}$$

and

$$\frac{dS_{nm}^{BA}}{dx} = \frac{\Delta h}{2}\frac{dD^{++B\quad A}}{dh}\left(2\frac{\alpha}{\pi} - 1\right) sin\left(\frac{\pi y}{L}\right) \tag{45}$$

$$\frac{dh}{dx} = \Delta h \cdot sin\left(\frac{\pi y}{L}\right)\left(2\frac{\alpha}{\pi} - 1\right) \quad (46)$$

respectively.

3.3 Computation and discussion

In the following computations, numerical solutions are obtained by computer software, such as Matlab, to quantitatively evaluate the resulted formulas for mode-conversion coefficients of VLF radio waves due to sunrise and sunset.

3.3.1 Mode-conversion coefficients of reflected and scattered vectors

During long-distance propagation across the transition regions, several modes contribute to the net fields. Based on the derived formulas, the vertical electric field component can be determined by the distance variations, so that the mode-conversion coefficients can be obtained, correspondingly. Consider that the altitude of the lower ionosphere is assumed to be at **90*km*** and **70*km*** for nighttime and daytime, respectively, and the radius of earth is taken by $\boldsymbol{a} = \mathbf{6370}\boldsymbol{km}$. In the computations, the ground can be characterized by the plane sea surface with conductivity and relative dielectric constant of it being $\boldsymbol{\sigma_g} = \mathbf{4}$ **S/m** and $\boldsymbol{\varepsilon_{rg}} = \mathbf{80}$, respectively. When both receiving antenna and radiating antenna are located on the ground surface, the resulted mode-conversion coefficients are computed and summarized in **Tables 1** and **2** for distinct operating frequencies at $\boldsymbol{f} = \mathbf{10}\boldsymbol{kHz}$ and at

Nth	Anisotropic ionosphere			
	Re (S_{1m}^{BA})	Im (S_{1m}^{BA})	Re (S_{1m}^{AA})	Im (S_{1m}^{AA})
1	1.0554	0.1188i	−1.5452	−0.0717i
2	−0.5785	−0.0165i	0.4090	0.0143i
3	0.2533	0.0071i	−0.1753	−0.0061i
4	−0.1415	−0.0039i	0.0974	0.0034i
5	0.0902	0.0025i	−0.0620	−0.0022i

Table 1.
The transmitted coefficients and scattered coefficients S_{1m}^{BA} and S_{1m}^{AA}, at f = 10 kHz.

Nth	Anisotropic ionosphere			
	Re (S_{1m}^{BA})	Im (S_{1m}^{BA})	Re (S_{1m}^{AA})	Im (S_{1m}^{AA})
1	0.9883	0.0014i	−0.0939	−0.0372i
2	0.0946	0.0080i	0.1188	0.0074i
3	−0.0167	−0.0038i	−0.0509	−0.0032i
4	0.0069	0.0021i	0.0283	0.0018i
5	−0.0038	−0.0014i	−0.0180	−0.0011i
6	0.0025	0.0009i	0.0125	0.0008i
7	−0.0017	−0.0007i	−0.0091	−0.0006i
8	0.0013	0.0005i	0.0070	0.0004

Table 2.
The transmitted coefficients and scattered coefficients S_{1m}^{BA} and S_{1m}^{AA}, at f= 20 kHz.

$f = 20kH$, respectively. The computation is obtained by choosing a nighttime-to-daytime propagation path with an abrupt effective height change of ionosphere by $\mathbf{20}\boldsymbol{km}$ across the discontinuity.

Technically, the magnitudes of the first-second and second-first mode-conversion coefficients will affect more than the higher modes [19]. It is seen that more summation terms would be required to guarantee accuracies of calculations at high operating frequencies. In **Tables 1** and **2**, the determined elements of the reflected and transmitted scattering matrix are calculated by Eq. (36) and Eq. (37), respectively. The derived matrix is subjected to modal equations by Eq. (8). It is noted that for more complicated case of varying ionosphere heights, these coefficients can be derived by Eq. (42) and Eq. (43) which are not addressed in the illustrative example for simplicity.

3.3.2 Interference behaviors of VLF radio waves across abrupt discontinuity

For a better understanding of the interference pattern by the use of the obtained formulas, the following computations are developed when the ionosphere layer is composed of a succession of two bounded layers, each with a distinct character. Assuming that the transmitting antenna is placed on the sea surface in region A, the magnitudes of components for VLF radio waves vary as functions of the propagation distances and observation heights, which is plotted in **Figure 8**. It is noted that the altitude $\boldsymbol{h}$ enters in the calculation of excitation factors and of field amplitudes; and $\boldsymbol{h}$ of the waveguide region around the source (section A in **Figure 5**) is left the same as in the original mode calculations [**Figure 8(a)** for nighttime $\boldsymbol{h_N} = \mathbf{90}\boldsymbol{km}$ and **Figure 8(b)** for daytime $\boldsymbol{h_D} = \mathbf{70}\boldsymbol{km}$, where the subscripts $\boldsymbol{N}$ and $\boldsymbol{D}$ refer to nighttime and daytime conditions, respectively]. The other waveguide region of the discontinuity problem does not contain a source [section $\boldsymbol{B}$ of **Figure 5(a)** and **Figure 5(b)**]; after determining the attenuation rates and phase velocities (or the transmitting scattering coefficients $\boldsymbol{S_{nm}^{DN}}$) for the waveguide modes of this source-free region with the calculated equivalent ionospheric surface impedance matrix to be discussed in the next section, the scattered fields in the waveguide in both sections due to sunrise can be calculated at any arbitrary altitude.

From **Figure 8(c)** and **Figure 8(d)**, the distributions of the electromagnetic field strength for VLF waves due to sunrise are depicted at the operating frequency f = 10 kHz. From **Figure 8(e)** to **Figure 8(g)**, it is depicted that the field components in the sections are quite different at different propagation distances and different altitudes. Obviously, the interference occurs at the abrupt junction of two cascaded waveguides.

In a similar manner, the wave components are evaluated through a D-to-N propagation path. Considering that all parameters are chosen as same as those in **Figure 8**, the spatial distribution of the scattered field is depicted in **Figure 9(d)** through a D-to-N propagation path. However, it is seen that the interference pattern is unlikely in agreement with reciprocity in which the section on the right side is subjected to different incident angles. This shall also be affected by the earth geomagnetic field inclination angle and the propagation angle. However, the proposed computational scheme is in advantage of providing a straightforward in calculation for nonreciprocal model. In **Figure 9(e)**, the electromagnetic field in the transition region by corresponding formulas is plotted versus the propagation distance in the same figure at the operating frequency $f = \mathbf{10}\boldsymbol{kHz}$, in daytime and nighttime, respectively. The interference mechanism by D-to-N propagation path is similar to that of N-to-D propagation path in **Figure 10(a)**, and the field strength is coherent near the junction.

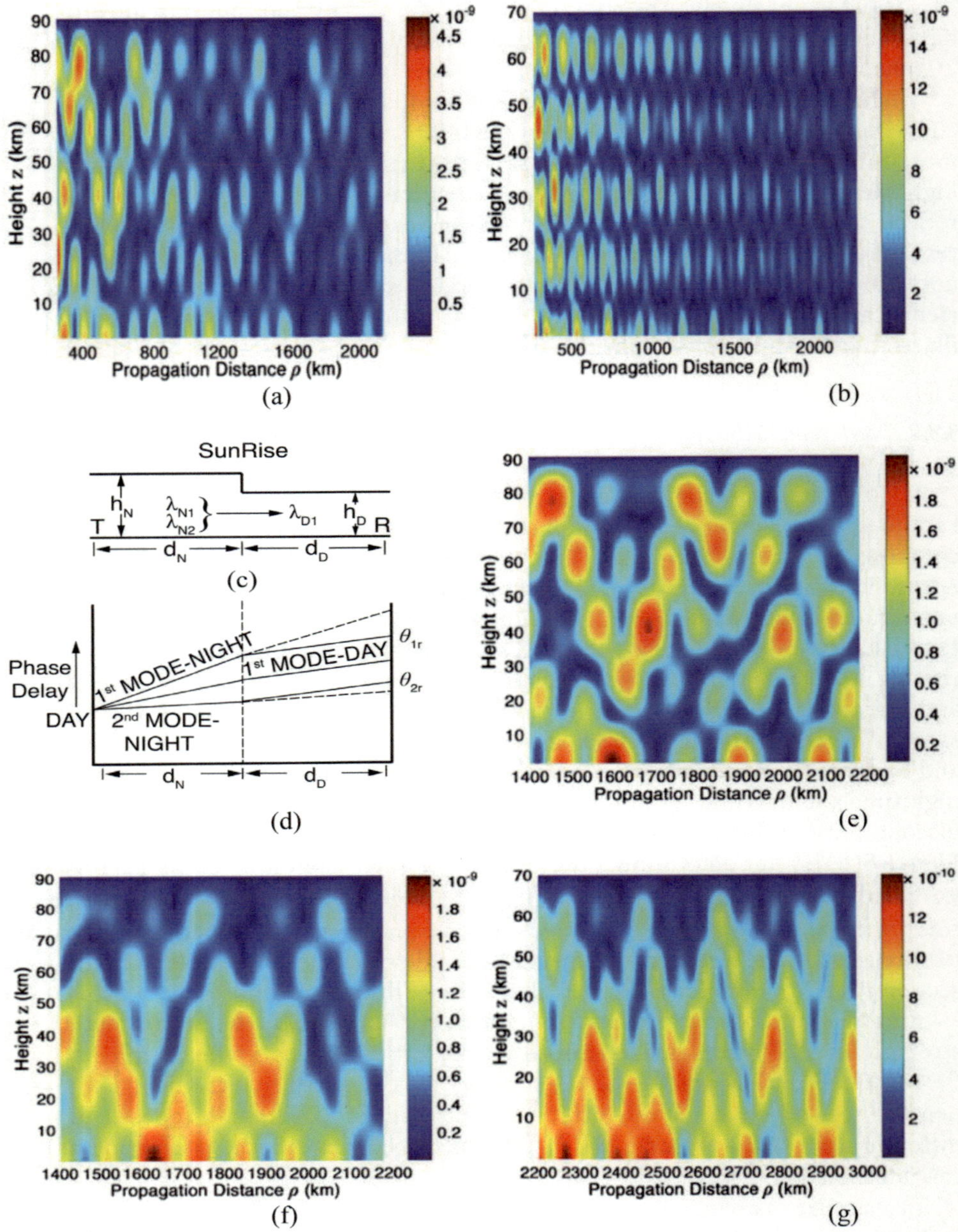

Figure 8.
Illustrative example of mode interferences in N-to-D propagation path with spatial distributions of the field component E_r *in* V/m, *at the operating frequency* f = 10 *kHz; with parameters taken by* $\varepsilon_g = 80$, $\sigma_g = 4$ *S/m*, $h_N = 90km$, *and* $h_D = 70km$; *with (a) incident field in nighttime, (b) incident field in daytime, (c) geometry of waveguide, (d) mode interference model, (e) incident field in region a, (f) scattered field in region a, and (g) transmitted field in region B.*

3.3.3 Comparing with the experimental data for the VLF radio wave at 20.5 kHz due to sunrise

In **Figure 10(a)**, the interfered electric field due to sunrise is computed as comparing with that for a homogeneous waveguide representing for daytime and nighttime, respectively. It is seen that the interference subjected to mode

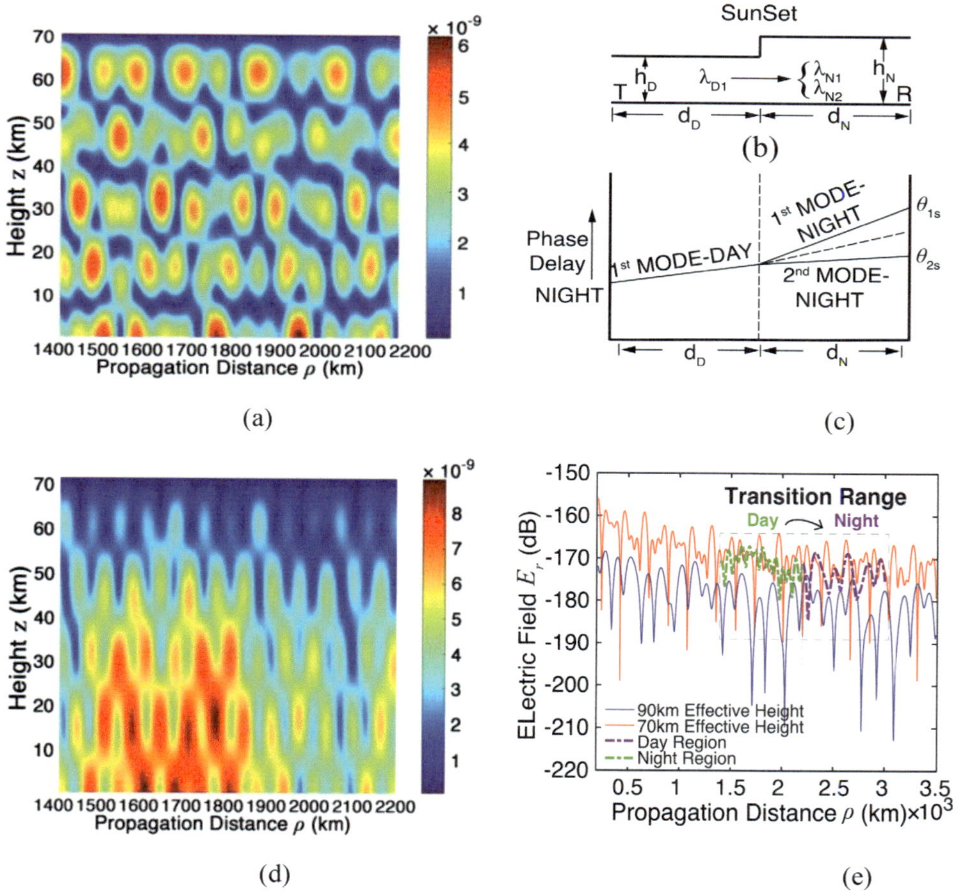

Figure 9.
Illustrative example of mode interferences in D-to-N *propagation path with spatial distributions of the field component* E_r *in* V/m, *at the operating frequency* f = 10 *kHz;* $\varepsilon_g = 80$, $\sigma_g = 4$ *S/m*, $h_N = 90$*km, and* $h_D = 70$*km, with (a) incident wave in region a, (b) geometry of waveguide, (c) mode interference model, (d) scattered wave in region a, and (e) the magnitudes of the field component* E_r *in dB versus the propagation distance due to sunset.*

conversion is resulted from discontinuousness from distinct ionosphere boundaries due to day and night propagation paths.

In order to validate of the proposed computational scheme, the simulation has been compared with the experimental data addressed in a recent paper [25] in **Figure 10(b)**. The radiation power is 250 kw, and the radiating device is chosen by un-directional antenna. The experimental data are measured at the operating frequency f = 20.5 kHz in the same location from the time 4 o'clock a.m. to 8 o'clock a.m. in east-to-west direction. Taking into account that both the observation point and transmitting antennas are located on the ground, with the same operating frequency, the conductivity and relative dielectric constant of sea water being σ_1 = 4 S/m and ε_{r1} = 80, the amplitudes of synthetic fields are computed along an idealized N-to-D transition propagation path. The synthetic fields contain two parts: the first section before a sunrise line (including the incident wave and scattered wave) and the second section across a sunrise line (standing for the transmitted wave).

It is shown from **Figure 10(b)** that the calculated result as synthetic field being interfered by multimode-conversion effect is closer to the experimental data measured on 4 o'clock and 5 o'clock than measured on 6, 7, and 8 o'clock data near the

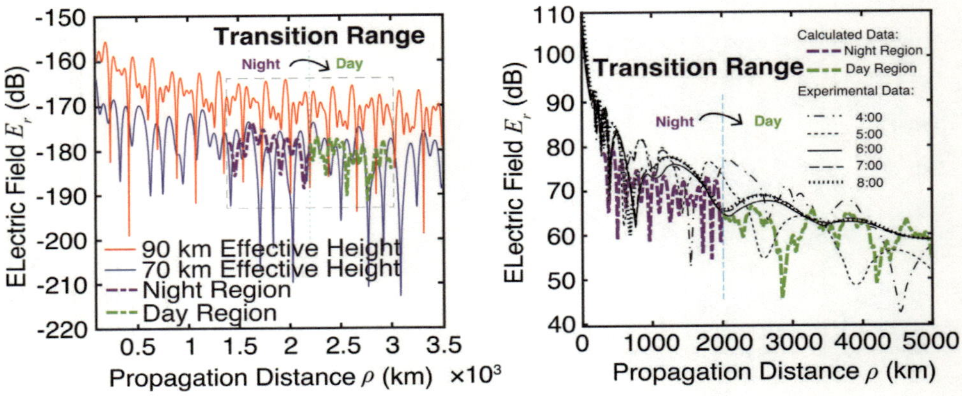

(a) (b)

Figure 10.
The magnitudes of the field component E_r in dB versus the propagation distance due to sunrise in nighttime, daytime, and transition range, respectively, at the operating frequency $f = 10$ kHz; $\varepsilon_g = 80$, $\sigma_g = 4$ S/m, $h_N = 90km$, and $h_D = 70km$, with (a) the simulated results comparing the transition path in N-to-D with daytime and nighttime propagations and (b) the calculated result comparing with experimental data, respectively.

sunrise line. Additionally, the measured data after 6 o'clock does not change too much since process of sunrise passed. Taking into account that the difference between calculated data and the measured data is resulted from the applied models and specific measuring conditions, such as the time, locations, and propagation paths, the proposed computational scheme would be more helpful by adjusting it into more practical models than a simple structure with abrupt discontinuity in height boundaries.

4. Conclusions

In this chapter, the effect of multimode interferences is investigated for VLF radio waves due to sunrise and sunset based on an idealized cascaded anisotropic inhomogeneous terrestrial waveguide. The coefficients of mode conversion for scattered vectors and transmitted vectors are evaluated at the junction of abrupt discontinuity representing the nighttime and daytime transition. The treatment provided can be enhanced in the future by a more practical terrestrial waveguide with varying effective ionosphere height rather than the abrupt one. The study may provide with some help in the applications of long-distance communication and navigation.

Acknowledgements

The authors are grateful to all referees for their constructive comments and suggestions in improving the quality of this chapter. The authors thank all associate editors very much for their help and encouragements.

Conflict of interest

No potential conflict of interest was reported by the authors.

Author details

Ting Ting Gu[1] and Hong Lei Xu[2]*

1 Department of Information Science and Electric Engineering, Zhejiang University, Hangzhou, China

2 Science and Technology on Electromagnetic Scattering Laboratory, Beijing, China

*Address all correspondence to: xhl_207@126.com

References

[1] Martin HG. Wave-guide mode propagation of very low frequency radio waves. Journal of Atmospheric and Terrestrial Physics. 1961;**20**:206-209. DOI: 10.1016/0021-9169(61)90078-2

[2] Rieker J. Sunset and sunrise in the ionosphere: Effects on the propagation of longwaves. Journal of Research of the National Bureau of Standards. 1963; **67D**:119. DOI: 10.6028/jres.067D.016

[3] Crombie DD. Periodic fading of VLF signals received over long paths during sunrise and sunset. Journal of Research of the National Bureau of Standards. 1964;**68D**:27-34. DOI: 10.6028/jres.068D.012

[4] Ries G. Results concerning the sunrise effect of VLF signals propagated over long paths. Radio Science. 1967;**2**: 531-538. DOI: 10.1002/rds196726531

[5] Smith R. Mode conversion coefficients in the earth-ionosphere waveguide for VLF propagation below a horizontally stratified, anisotropic ionosphere. Journal of Atmospheric and Terrestrial Physics. 1977;**39**:539-543. DOI: 10.1016/0021-9169(77)90163-5

[6] Devi MI, Khan I, Madhusudhana RDN. A study of VLF wave propagation characteristics in the earth-ionosphere waveguide. Earth, Planets and Space. 2008;**60**:737-741

[7] Steele FK, Crombie DD. Frequency dependence of VLF fading at sunrise. Radio Science. 2015;**2**:547-549. DOI: 10.1002/rds196726547

[8] Walter F, Scarabucci RR. VLF propagation in the magnetosphere during sunrise and sunset hours. Radio Science. 1975;**10**:965-971. DOI: 10.1029/rs010i011p00965

[9] Rodger CJ, Brundell JB, Dowden RL, Thomson NR. Location accuracy of long distance VLF lightning location network. Annales Geophysicae. 2004; **22**:747-758. DOI: 10.5194/angeo-22-747-2004

[10] Bennett AJ, Gaffard C, Nash J, et al. The effect of modal interference on VLF long-range lightning location networks using the waveform correlation technique. Journal of Atmospheric and Oceanic Technology. 2011;**28**:993-1006. DOI: 10.1175/2011JTECHA1527.1

[11] Wait JR. Reflection of VLF radio waves at a junction in the earth-ionosphere waveguide. IEEE Transactions on Electromagnetic Compatibility. 1922;**34**:4-8. DOI: 10.1109/15.121660

[12] Wait JR. Mode conversion and refraction effects in the earth-ionosphere waveguide for VLF radio waves. Journal of Geophysical Research. 1968;**73**:3537-3548. DOI: 10.1029/ja073i017p05809-01

[13] Bahar E. Analysis of mode conversion in waveguide transition sections with surface impedance boundaries applied to VLF radio propagation. IEEE Transactions on Antennas and Propagation. 1968;**16**: 673-678. DOI: 10.1109/TAP.1968.1139274

[14] Bahar E. Propagation of radio waves in a model nonuniform terrestrial waveguide. Proceedings of the Institution of Electrical Engineers. 1966;**113**: 1741-1750. DOI: 10.1049/piee.1966.0299

[15] Bahar E. Scattering of VLF radio waves in the curved earth-ionosphere waveguide. Radio Science. 1968;**3**: 145-154. DOI: 10.1002/rds196832145

[16] Bahar E. Propagation of VLF radio waves in a model earth-ionosphere waveguide of arbitrary height and finite surface impedance boundary: Theory

and experiment. Radio Science. 1966;**1**: 925-938. DOI: 10.1109/APS.1965.1150285

[17] Galejs J. Propagation of VLF waves below a curved and stratified anisotropic ionosphere. Journal of Geophysical Research. 1964;**69**:3639-3650. DOI: 10.1029 /jz069i017p03639

[18] Galejs J. VLF propagation across discontinuous daytime-to-nighttime transitions in anisotropic terrestrial waveguide. IEEE Transactions on Antennas and Propagation. 1971;**19**: 756-762. DOI: 10.1109/TAP.1971.1140034

[19] Pan WY, Li K. Excitation and propagation of SLF/ELF electromagnetic waves in the earth-ionosphere waveguide/cavity. In: Pan WY, Li K, editors. Propagation of SLF/ELF Electromagnetic Waves. 1st ed. Hangzhou, CN: Zhejiang University Press; Berlin Heidelberg, GER: Springer-Verlag; 2014. pp. 17-63

[20] Li K. Electromagnetic Fields in Stratified Media. Hangzhou, CN: Zhejiang University Press; Berlin Heidelberg, GER: Springer-Verlag; 2009

[21] Li GZ, Gu TT, Li K. SLF/ELF electromagnetic field of a horizontal dipole in the presence of an anisotropic earth-ionosphere cavity. Applied computational electromagnetics. Society. 2014;**29**:1102-1111

[22] Li K, Sun XY, Zhai HT. Propagation of ELF electromagnetic waves in the lower ionosphere. IEEE Transactions on Antennas and Propagation. 2011;**59**: 661-666. DOI: 10.1109/TAP.2010.2096391

[23] Wait JR. Electromagnetic Waves in Stratified Media. 2nd ed. New York: Pergamon Press; 1970

[24] Gu TT, Xu HL, Li K. Mode interferences of VLF waves in an anisotropic waveguide due to sunrise and sunset. IEEE Transactions on Antennas and Propagation. 2018:1-1. DOI: 10.1109/TAP.2018.2870347

[25] Wei L, Jiang YZ, Li KF. VLF field strength calculation methods based on waveguide mode. Ship Electronic Engineering. 2011;**59**:661-666

Section 5

Physics of Waveguiding

Chapter 9

Radiation Fluxes Waveguide-Resonance Phenomenon Discovered in Result of X-Ray Nanosize Beam Formation Study

Egorov Evgenii Vladimirovich and
Egorov Vladimir Konstantinovich

Abstract

The work discussed shortly the experimental results, which was the waveguide-resonance mechanism relevation forerunner of characteristic X-ray radiation flux propagation. Technology of the planar air extended slit clearance preparation is presented. The methodology of X-ray beam parameter study formed by these slit clearances, which allowed to find the critical parameter answering for the radiation flux propagation mechanism change from the multiple external total reflections to the waveguide-resonance one, is described. Main features of the X-ray flux waveguide-resonance propagation mechanism were revealed. The self-consistent model of the mechanism is displayed with details. It is shown that the waveguide-resonance effect has universal character, and it reflects the fundamental nature phenomenon. The peculiarities of X-ray device functioned in frame of the phenomenon manifestation planar X-ray waveguide resonator (PXWR) and the increasing methods of its practical efficiency are discussed. The phenomenon practical application is presented concisely.

Keywords: X-ray flux, external total reflection, X-ray standing wave, coherence length, X-ray nanophotonics, planar X-ray waveguide resonator, waveguide-resonance propagation phenomenon, spatial coherence, angular divergence, partial angular tunneling effect

1. Introduction

The problem of X-ray beam formation with a minimal size cross section and a small angular divergence is the central problem of all X-ray diagnostical methods. The first real step for the solution of this problem was connected with the names of P. Hirsch and J. Keller suggested to form X-ray microbeams by employing the glass capillary [1]. More recently, the planar thin film waveguides have been offered for X-ray microbeam formation [2]. Authors of the work showed that the waveguide with a material media core for X-ray beam transportation can form small size beams, but the beam intensity attenuation was very great. In the direction, development similar investigations were carried out in a number of experimental works

[3–5]. Authors of these investigations have managed to obtain X-ray beams with a width of 100 nm, a height of some millimeters, and the total intensity near 5×10^7 photon/s in condition of the resonant synchrotron radiation coupling. The significant progress in these research works was achieved by switching over a study from coupling mechanism of the emergent beam preparation to the ones based on the radiation transportation by the core layer from input of the waveguide to its outlet. In result of the flux mode structure analysis, the phenomenon of an X-ray standing wave arising was mentioned [4]. The properties of X-ray beams formed by polycapillary optics systems have been intensively studied, too [6–10]. The optics of systems is based on the phenomenon of X-ray beam multiple total external reflections on the inner surface of a quartz capillary. Mono and polycapillary optics are the beautiful facilities for the formation of microsize beams. At the same time, the polycapillary optics is characterized by significant losses of X-ray beam intensity in the transportation process. The problem of X-ray flux intensity losses was the subject of a specific investigation [11]. Authors of the work studied the effect of capillary damages in result of X-ray beam influence. They demonstrated the linear worsening of X-ray beam transmission ability for the glass capillary with an increase of radiation dose. This effect is not significant for the quartz guides of X-ray fluxes.

In parallel with the traditional approaches to micron and submicron X-ray beam formation mentioned above, it has been discovered the specific technique of a superfine beam preparation by using the so-called "slitless" collimator [12–14] formed by two quartz plane polished plates mated together. Its device lets to form X-ray emergent beam with the visible magnitude of a radiation intensity compared with the incident beam intensity value [12]. Unfortunately, the study of this phenomenon and attempts of its practical application have been undertaken in recent years only [15, 16]. In fact, the slitless X-ray collimator represents the planar waveguide with a minimum size of an air slit. Width of the slit is defined by roughness and waviness levels of the collimator reflector plane surfaces. At the same time, the air core between guide claddings is the ideal waveguide channel from standpoint of the radiation flux intensity preservation. Similar waveguides with fixed and tunable air gaps have begun to find the practical application in the works of Zwanenburg group [17, 18]. Their waveguides with Cr claddings and air core can produce the emergent beam with a width of d = 500 nm, a height of h = 0.1 mm, and a total intensity of $J = 2.4 \times 10^7$ photon/s [17]. Great omission of these works consisted in the ignoration of X-ray standing wave arising in air core of their radiation guides.

The original glance on the problem of X-ray flux transportation by a planar extended slit clearance was presented in the works of Kawai group [19]. As opposite to the standing wave conception, authors included the specific notion about X-ray traveling waves or Yoneda wing. This approach has some grounding in theory [20, 21]. But these works left behind bracket the interference interaction between falling and reflecting fluxes.

Very strange approach was suggested by Dabagov for the description of X-ray flux transportation by a hollow quartz capillary [22]. Instead of the conventional conception connected with the multiple total external reflection mechanisms, author advanced the idea of "X-ray quantum subsurface channeling." We believe that the approach is not pragmatic since the channeling phenomenon offers a photon motion in the periodic potential, but the surface of amorphous quartz cannot produce the correct periodic field.

A number of publications with model description attempts of X-ray flux propagation through a narrow extended slit are presented in the literature [23–26]. These models are built on the working hypothesis that X-ray radiation is the planar monochromatic electromagnetic wave. But it is universally known that the realistic X-ray sources produce the quasimonochromatic radiation fluxes with λ_0 average wavelength and $\Delta\lambda$ monochromatism degree.

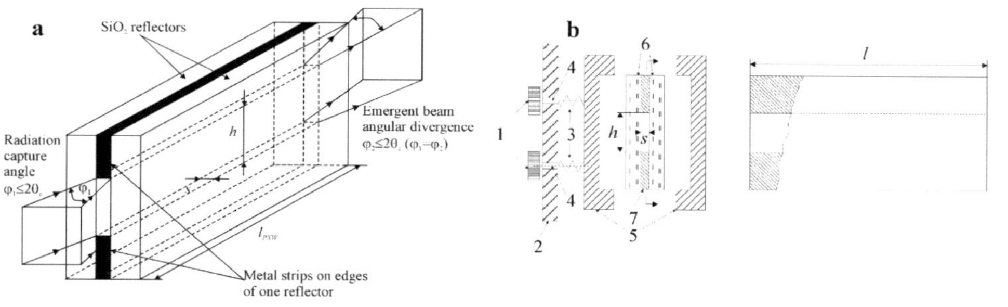

Figure 1.
Scheme of X-ray initial flux capture area and the emergent beam formation by the planar extended slit clearance formed by two quartz reflectors (a) and real construction of the device for the study of the spatial intensity distribution dependence in X-ray beams on the slit clearance width (b). 1: aligning handles; 2: installation plate; 3: spring pawls; 4: fine tuning screws; 5: radiation guide holder; 6: quartz reflectors with length 100 mm; h and s: height and width of slit clearance ($\varphi_1 = \varphi_2$).

Materials with information about Δλ magnitudes featured for X-ray characteristic radiation produced by X-ray laboratory sources are presented in Handbook editions [27, 28]. But the main shortage at interpretation of X-ray flux transportation by different waveguide structures including slitless collimator devices was the statement that X-ray flux propagation takes place accordingly to the multiple total external reflection mechanisms as sole possible one. We were skeptical of this point of view and decided to produce the systematic investigation of the planar extended slit clearance width influence on its X-ray emergent beam parameters. For similar investigations, we selected construction presented in **Figure 1**. It is the air planar extended X-ray waveguide (PXW).

2. Technological features of PXW fabrication

The main components of the PXW structure are planar polished dielectric reflectors forming its radiation-transporting air slit clearance. In preparation of reflector working surfaces, it is necessary to fulfill a number of technological requirements guaranteeing the desired surface quality. The technical parameters that determine the surface quality are first of all the roughness and waviness and, moreover, a specific factor associated with local work hardening arising from nonuniform surface heating during polishing.

Modern polishing methods are capable of ensuring a surface roughness level of about 0.5 nm. Such a high degree of polishing can be controlled via direct testing with the help of atomic-force microscopy. The aforementioned roughness level is quite comparable with the range within which the potential on the condensed material surface varies from the value typical of its entire volume to that corresponding to vacuum [29, 30]. At the same time, the atomic-force microscopy technique makes it impossible to estimate the surface waviness and, moreover, the level and degree of surface distortions caused by the appearance of local work hardening. To a certain extent, the influence of these parameters on the surface quality can be estimated with the help of an optical method based on violated total internal reflection [31], which enables us to discard reflectors with appreciable contributions to the deterioration of the reflector surface quality.

In preparation of waveguide resonators, the most critical technological stage is thin film metallic strip deposition on the edges of one of the quartz reflectors constituting a pair used to create a waveguide-resonance channel. The deposited materials are titanium or chromium with a high degree of adhesion to the quartz surface. During the deposition process, the surface of the future waveguide-resonance channel was coated with aluminum foil. Thin film metallic strips were primarily deposited in the vacuum chamber of a Leybord LG L-560 setup via the electron beam evaporation method. The film coating growth rate was 0.1 nm/s. During the deposition process, the chamber pressure was maintained at a level of 10^{-4} Pa. However, in spite of relatively high vacuum, the metallic-strip material contained a certain number of oxygen atoms (up to 10 at%). When the films were deposited, some reflectors were heated up to 80°C. As a result, the density of coating adhesion to the quartz glass surface increased appreciably. A simplified diagram of the mutual arrangement of assemblies in the chamber used to the deposit coatings in vacuum is depicted in **Figure 2**. The position of the reflector intended for coating deposition is symmetric with respect to the point source of metal atoms.

The basic requirement to the quality of the prepared strip coatings is thickness homogeneity along the entire length of the PXW reflector. Let us consider the geometry of the diagram, as shown in **Figure 2**. Then, under the assumption of angular homogeneity of the metal atom flux excited by the electron beam, it can be expected that the deposited strips will be characterized by a nonuniform coating thickness and its largest value will be at the reflector center. For coating deposition condition optimization, it is necessary to employ the thickness control methods. At the center (t_2) and edges (t_1) of the reflector (**Figure 3**), the deposited strip thicknesses were determined via the Rutherford backscattering (RBS) of He^+ ions with the help of "blank samples." Single-crystal silicon samples located on aluminum foil, which covered the surface during deposition, were used as these blank samples. Thus, each reflector with deposited metallic coatings can be characterized by at least two Ti/Si blank samples. Their experimental investigations were performed by means of the Sokol-3 ion beam analytical complex situated at the Institute of Microelectronics Technology and High Purity Materials, Russian Academy of Sciences [32]. The results of these measurements are depicted in **Figure 4**.

Using the spectra of the RBS of He^+ ions (E_0 = 1 MeV) (**Figure 4**), it is possible to perform accurate determination of the thicknesses of the strips deposited on reflectors in their central part (**Figure 4a**) and on the edges (**Figure 4b**). These thicknesses are determined by approximating peaks with almost flat tops, which

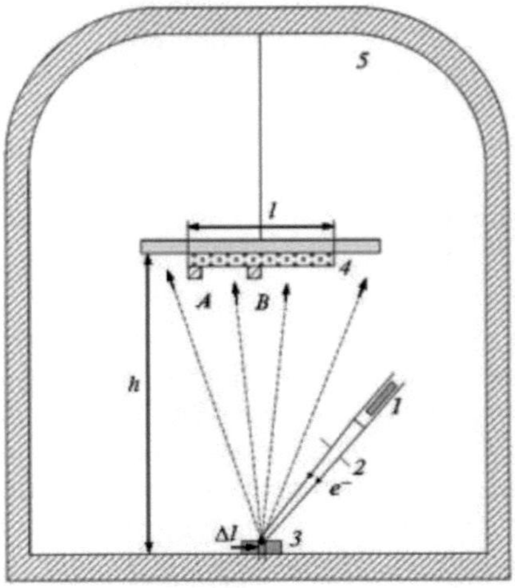

Figure 2.
Simplified diagram of the chamber used to deposit titanium strips on quartz reflectors: (1) electron gun, (2) focusing system, (3) titanium target, (4) quartz reflector, (5) vacuum volume, h = 200 *mm,* l = 100 *mm; and* A *and* B *are the positions of blank samples during deposition.*

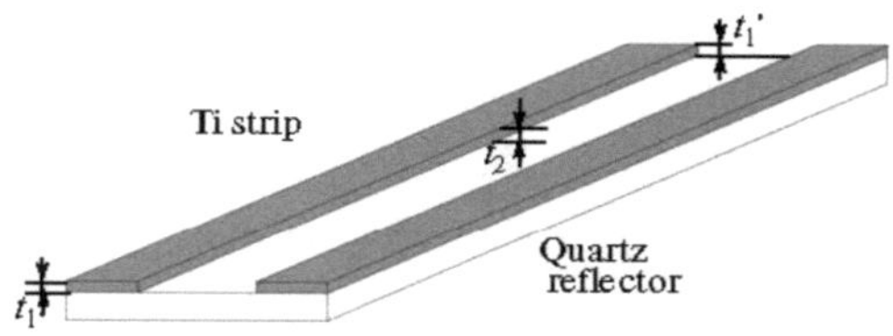

Figure 3.
Waveguide-resonator reflector with thin film strips on the edges: t_1 and are the coating thicknesses on its edges, and t_2 is the thickness in the central part.

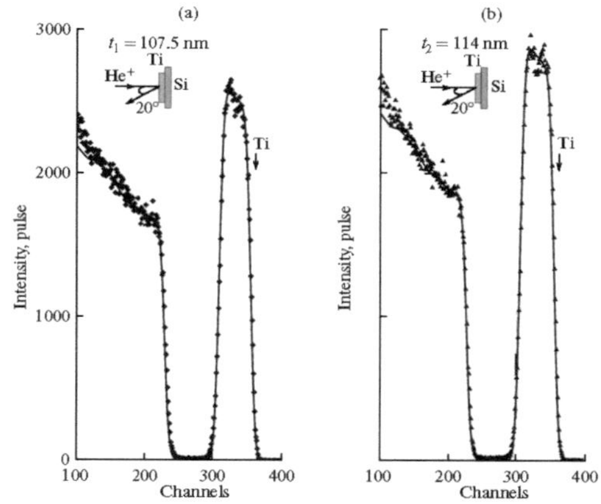

Figure 4.
Typical spectra of the RBS of He⁺ ions (E_0 = 1 MeV). Data were obtained for Ti/Si blank samples located at the positions A *and* B *(**Figure 2**) corresponding to the (a) edge and (b) center of the strip coatings of the quartz reflectors.*

correspond to He^+ ion scattering from the coating atoms. (The low energy steps observed in the RBS spectra of the tested targets correspond to He^+ ion scattering from substrate atoms.) The presented spectra were mathematically processed using RUMPP, the modified version of the famous RUMP approximation program [33]. Approximation of the aforementioned spectra indicates that the film strip thicknesses are t_2 = 114 ± 0.5 nm in the central part and t_1 = 107.5 ± 0.5 nm on the edges. Thus, the inhomogeneity in the thickness of the strips deposited onto the given reflector is 6%. This result completely coincides with the estimate based on geometric considerations. The above data correspond to the reflector whose coatings were deposited at a distance of h = 200 mm between the source of the evaporated atoms and its surface (**Figure 2**). The reflector length is l = 100 mm. Under the assumption that the deposition rate is proportional to the squared distance from the source, the expected difference turns out to be 6.1%. For the coating thickness difference decreasing, we increased the h-distance up to 1000 mm. In that case, the thickness difference accordingly to RBS data achieved to 1%. Experiments showed that the similar conditions are acceptable for X-ray waveguide assemblage.

Specific attention has been given to the direct determination of the effective slit width in different waveguides and in a slitless collimator because the data presented in early works about slitless devices [12–14] were not clear with respect to the width. The width was evaluated by very effective optical method connected with the attenuated total internal reflection effect [31]. In our investigation, we used the laser source with λ_0 = 680 nm. **Figure 5** presents the measurement geometry. The studied waveguide was situated in a specific cartridge equipped by black light absorber. The light beam introduced into the waveguide by using the quartz prism

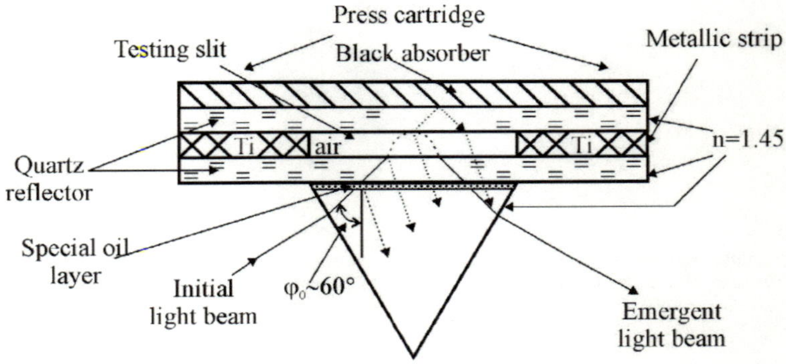

Figure 5.
Principle scheme for direct measurements of a waveguide slit width by methods of the attenuated internal total reflection (AITR). I and II are quartz reflectors of a waveguide. Scheme was published, in first, in Ref. [34].

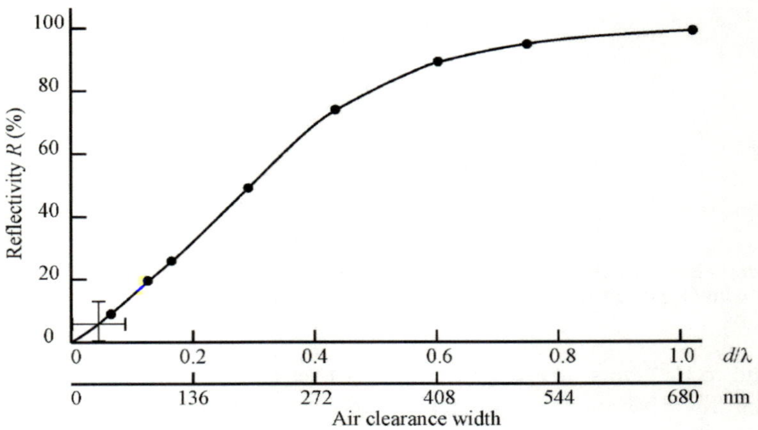

Figure 6.
Experimental data presented the relationship between the waveguide slit width and the reflectivity magnitude are obtained by AITR method. The point characterized the slitless collimator has a specific design.

fixed on the waveguide reflector by specific oil (n = 1.45). The prism could change its position on the reflector surface. Emergent light beam in the measurement process registered by standard photodiode equipped by circular aperture with a diameter of d = 0.5 mm. In the measurements process, the light beam in transit through the prism incidented on the waveguide, underwent the attenuated total internal reflection on the waveguide slit clearance and a lux meter recorded the reflection intensity. The normalization measurement was executed by using the waveguide with a width slit of s = 0.12 mm. Specific details of investigations are described elsewhere [34]. The measured data for several waveguides with different slit widths are presented in **Figure 6**.

Standard least square method was used for the experimental data fitting allowed to get a relationship between the light beam reflectivity factor and the width of waveguide slit clearance. In process of the slitless collimator study, we registered the gap width variation in interval 0–60 nm at the prism translation along the slitless unit. In result, we concluded that the slitless collimator is characterized by effective width of the gap s = 30 $\pm$ 30 nm.

3. Experimental setup for the radiation intensity distribution study

The main device to study the X-ray intensity distribution was the HZG-4 diffractometer manufactured by Carl Zeiss Jena Firm. We produced some modification of the device by its detector circle radius increasing up to 500 mm. In the modification result, the measurement space resolution improved in three times. The measurement spectroscopic circuit was completed by NIM standard units produced by Ortec firm. The shaping time of amplifier unit was selected as 0.5 μs. Such selection allowed to get the pulse registration count rate up to 100 kHz. The design of our registration setup is presented in **Figure** 7. X-ray diffractometer used as the setup background is characterized by scanning regimes in nonstop function and start-stop moving with a minimum step of $\delta(2\theta) = 0.001°$. X-ray detector was equipped by slit-cut arrester with a width of $s = 0.1$ mm and a height of $h = 10$ mm and Soller slit system limiting the registered flux vertical divergence by value near 2°. X-ray flux take-off angle was selected near 6°. Main volume of experimental investigations was executed by X-ray tube BSW-24 (Cu) in regime $U = 20$ keV and $I = 10$ mA. Similar tube with Fe anode was exploited in some selective measurements. The X-ray space intensity distribution data collection was produced with the use of Cu filter attenuator characterized by the CuKα radiation decreasing factor $K = 200$. For the energy spectrum characterization, our facility setup was equipped by a pulse multichannel analyzer ACCUSPEC Canberra Packard in the form of PC computer board. In measurements of PXW parameters, we used the characteristic part of initial X-ray spectra only.

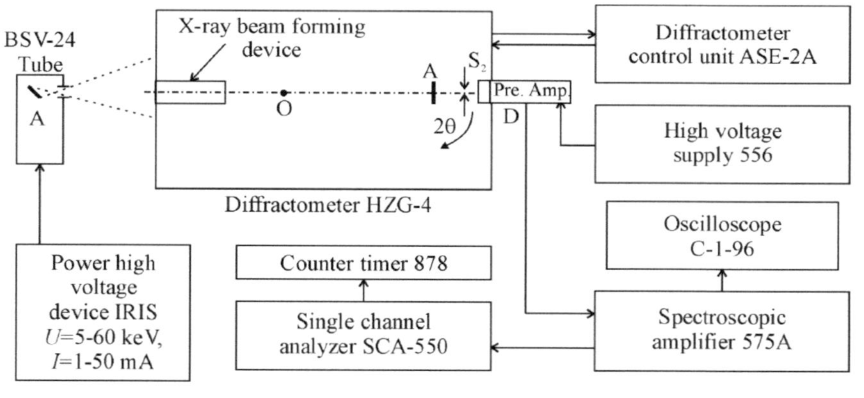

Figure 7.
Instrumental facility for the spatial distribution study of a quasimonochromatic radiation intensity in X-ray beams formed by quartz planar waveguides.

4. Angular radiation intensity distribution in X-ray beams

In the course of our measurements, the waveguide position in experimental setup in experimental process was not changed. In the experimental process, the distance between the waveguide inlet and the X-ray tube focal position was 75 mm, and the distance between the waveguide outlet and the X-ray detector slit was 460 mm. X-ray flux capture angle calculated on the basis of geometric approach was equal to 0.08° owing to the size of tube focus projection evaluated as 0.1 mm. In experiments, the diffractometer angular step $\Delta(2\theta)$ was 0.02°. At the same time, the detector slit angular acceptance was 0.01°. The single channel analyzer during experiments transmitted only pulses connected with the Cu characteristic radiation. Scheme of experimental measurement is presented in top position of **Figure 8**. Experimental results are shown in bottom of **Figure 8**.

The slit clearance size interval $0 \leq s \leq 2$ μm was characterized by the absence of the intensity distribution profile transformations for waveguide emergent X-ray beams. Profiles of these distributions were beautifully described by the Gaussian function. It showed some differences in FWHM values and total intensity magnitudes. The intensity magnitude scattering is likely due to our polishing technology wretchedness, but its increasing at $s > 200$ nm is higher as the experimental error. The distribution FWHM was not exceeded the radiation capture angle. At the same time, when the slit clearance width had exceeded 3 μm, the space intensity distribution found new form, which could be interpreted as a set of lines. Experimental data of the waveguide emergent beam total integral intensity dependence on the slit clearance width are presented in **Figure 8**. These dependence can be described by three typical size interval: $s \leq 200$ nm (a), $200 \leq s \leq 3000$ nm (b), and $s \geq 3$ μm (c) with different characters of the dependence.

Registered experimental data and common sense allowed to assume that the X-ray waveguide emergent beam consists of some independent deposits. One can suppose by using the geometrical optics concept that one of them is connected with the X-ray beam direct propagation through the waveguide slit clearance without

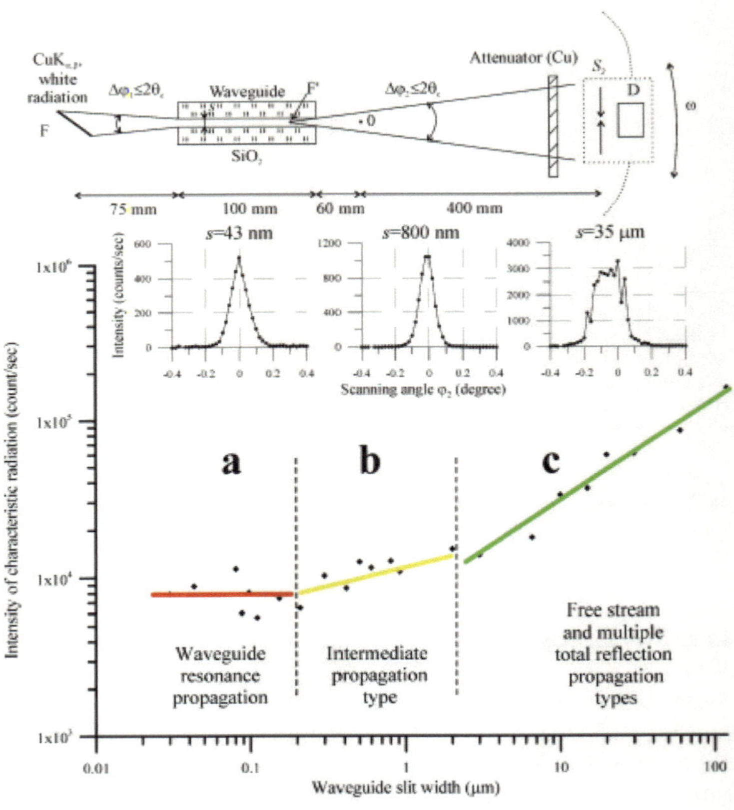

Figure 8.
Experimental scheme for the spatial distribution study of X-ray beam intensity formed by PXWs (1) patterns of $CuK_{\alpha\beta}$ intensity spatial distribution in beams formed by PXWs with slit sizes 43 nm, 800 nm, and 35 μm (2), and the experimental dependence of $CuK_{\alpha\beta}$ total intensity in beams formed by PXWs on its slit width (3).

waveguide reflector surface interaction. It is clear that at any width of the slit clearance, the direct beam will form its own partial peak and will show the linear dependence of its integral intensity variation on the slit clearance width. Its intensity will be equal to zero for the case of the slit clearance zero magnitude. Experimental data showed that the direct beam propagation mechanism was not able to describe the integral beam intensity variation dependence on the slit clearance width, especially, for the nanosize slit interval (a). In this region, the emergent beam total intensity maintains constant magnitude, which is more higher than zero. Calculations showed that the direct beam deposit into the experimental data magnitude for this region is less that 1%. Second region (b) is characterized by the monotonous integral intensity increasing at growth of the slit clearance width. This effect can be connected with a deposit increasing the direct beam in the total X-ray beam intensity. Third region (c) defined as $s > 3$ µm demonstrates sharp intensity growth at the width increasing. This area is characterized by an addition deposit appearing connected with X-ray beam multiple total external reflections on waveguide reflector surfaces. The slit clearance width increasing in this region leads to linear growth of intensity deposits defined by mechanisms of X-ray beam direct and multiple external total reflection propagations. The total integral intensity growth of the waveguide emergent beam break off when the slit clearance width exceeds size of the radiation source focus projection.

The experimental data comparison featured for different regions of the slit clearance width and peculiarities discussion of different mechanisms of radiation fluxes propagation insist us on conclusion that the nanosize region (a) is characterized by the specific waveguide-resonance manner of X-ray flux propagation [35]. Devices functioned in frame of the resonance manner we called the planar X-ray waveguide resonators (PXWRs) [29]. PXWR forms the X-ray quasimonochromatic flux as the indivisible ensemble with parameters, which are not depended from the slit clearance width and the initial distribution in radiation flux captured by the device. The waveguide mechanism of the X-ray quasimonochromatic flux propagation featured for the narrow extended slit clearance demonstrates the X-ray radiation density increasing and decreasing the irreversible losses. Angular divergence of PXWR emergent beam is equal to its radiation capture angle, and they cannot exceed twice value of the total reflection critical angle featured for the reflector material.

The slit clearance width intermediate interval (region b) is connected with two independent deposits into integral intensity defined by direct and quasiresonance beam propagation mechanisms. The increase of slit clearance ensures small growth of the beam integral intensity, but its radiation density diminishes. Spatial intensity distribution featured for this region shows a single-component form owing to small influence of the multiple total external effects on the emergent beam integral intensity. This effect deposit into the intensity becomes decisive factor when the slit clearance width exceeds critical value $s = 3$ µm (c). In this case, emergent beam divergence arrives its maximum $\Delta\theta = 2\theta_c$ independently from the device input aperture magnitude. The intensity distribution demonstrates multicomponent structure.

With practical point of view, it is very interesting to compare the radiation density parameter featured for PXWR and waveguides corresponding to "b" and "c" regions with similar parameter featured for X-ray beams formed by a conventional slit-cut device. The beam integral intensity on the slit-cut former output is more higher than one formed by different PXWs. But in the radiation density parameter, the planar extended waveguide structures are more effective. Direct comparison of the slit width is presented in **Figure 9**. Enhanced radiation density peculiar to X-ray beams formed by PXW is connected with width difference of the

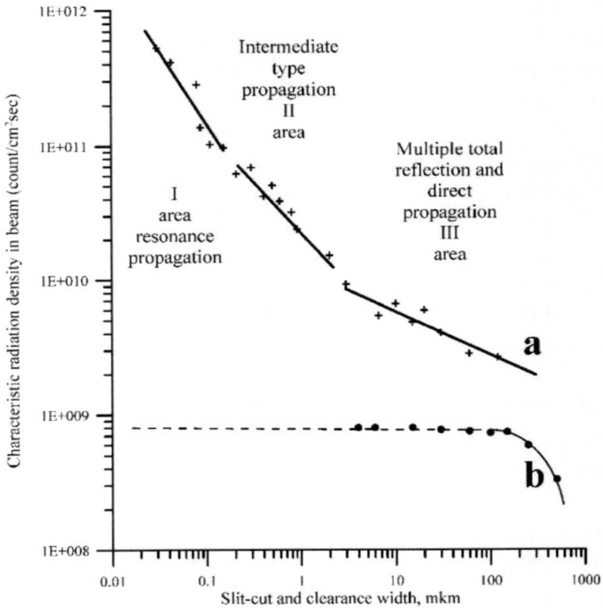

Figure 9.
Experimental dependences of CuKαβ flux radiation densities on slit width for X-ray beams formed by PXWs (a) and by the single slit-cut system (b).

waveguide slit clearance and the radiation source focus projection. Waveguides capture X-ray radiation fluxes in the angular aperture $\Delta\varphi \leq 2\theta_c$ from radiation source focus projection with width $P \sim 0.1$ mm into the slit clearance with more smaller width. In result, waveguide devices concentrate radiation. According to the data presented in **Figure 9**, PXWR is able to increase X-ray radiation density on three orders in its emergent beam in comparison with beams formed by slit-cut system. It is interesting that the maximum radiation density is expected for the slitless collimator. But the practical use of X-ray slitless system is troubled owing to the absence of the intensity stationary in its emergent beams.

5. Waveguide-resonance model for X-ray flux propagation

Figure 10a presents the idealizing scheme of X-ray flux total external reflection, which takes into account the degree of a radiation monochromatization $\Delta\lambda$ along

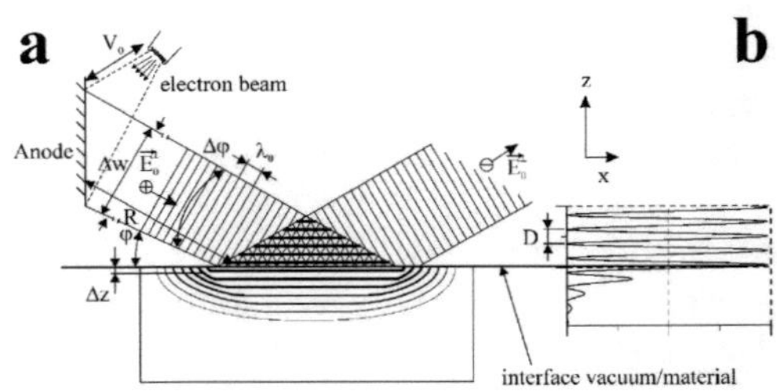

Figure 10.
Scheme of the interference field of X-ray standing wave arising from the external total reflection phenomenon in case of the quasimonochromatic flux interaction with material interface (a). Δw is the source focus projection, R is the distance between X-ray source and target position, φ is the angular flux divergence, Δz is the penetration depth, and D is the standing wave period. Standing wave intensities in the air area and in the reflector volume (b).

with the average wavelength λ_0. Owing to this principal limitation, the size of radiation standing wave area in the space over reflector will be bounded. $\Delta\lambda$ through the coherence length parameter characterizes the length of an electromagnetic radiation train or the photon longitudinal size. The interference phenomenon is possible if the path difference between the incident and the reflected fluxes does not exceed the magnitude of this parameter. But in any case, the longitudinal size of the interference area cannot exceed of the value. By this, it means that the coherence length of quasimonochromatic radiation is responsible for the longitudinal magnitude evaluation of X-ray standing wave area. In this framework of the phenomenological model, we accepted that the transverse size of the area is approximately equal to the longitudinal one. This premise is absolutely right so far as the real interference effect is connected with the spatial coherence of the quasimonochromatic flux radiation [36].

The next model postulate says that the description of the total X-ray reflection phenomenon must take into account the fundamental principle of a field continuity [37]. According to this principle, the interference field of X-ray standing wave cannot abruptly terminate on the material-vacuum (air) interface. The conventional model of X-ray beam total external reflection suggests that the radiation electromagnetic field amplitude undergoes exponential attenuation in the reflector material [38]. But the principle of electromagnetic field continually demands the exponential low multiplication on the interference term. Visualization of the modification is presented in **Figure 10b** and can be defined by the expression [39]:

$$\vec{E}_T(z,x,t) = \left[\frac{2\theta}{\theta + a + ib} e^{\frac{2\pi i z(a+ib)}{\lambda_0}}\right] e^{i\left(\omega_0 t - 2\pi p_x x\right)} \vec{E}_0 \tag{1}$$

where θ is the incident angle; λ_0 and ω_0 are the wavelength and the angular frequency of the radiation, respectively; p_x is the x-component of the photon momentum; and a and b are presented by the specific expressions [38]:

$$\left.\begin{aligned} a^2 &= \frac{1}{2}\left[\sqrt{\left(\theta^2 - 2\delta\right)^2 + 4\beta^2} - \left(\theta^2 - 2\delta\right)\right] \\ b^2 &= \frac{1}{2}\left[\sqrt{\left(\theta^2 - 2\delta\right)^2 + 4\beta^2} - \left(\theta^2 - 2\delta\right)\right] \end{aligned}\right\} \tag{2}$$

where δ and β factors are the formal parameters incoming into the conventional expression for the material refractive index [38]:

$$n = 1 - \delta - i\beta \tag{3}$$

The δ-factor is connected with the volume material polarization effect, and β-factor characterizes the attenuation degree of X-ray radiation flux in the material. But we would like to notice that the refractive index introduction in form presented by expression (3) is not correct, in principle. It suggests that the X-ray beam propagation velocity in material volume is higher than the velocity of the electromagnetic wave propagation in vacuum. Professor L.D. Landau at first pointed on this collision [40].

Figure 10a and **b** displays the principle model for an electromagnetic field distribution in the reflection area over and under the interface. The size of the interference area appeared over the interface is limited by the coherence radiation condition. But the interference area size under the interface is not limited. The entire volume of the reflector will be excited as a result of a flux total reflection on its local spot.

The external total reflection phenomenon is accompanied by an additional phase shift $\Delta\psi$ [41]. This parameter is the function of the flux incident angle φ. At the critical total reflection angle ($\varphi \cong \theta_c$), the additional phase shift strives to zero, but at the sliding incidence ($\varphi \cong 0$), $\Delta\psi$ value approximates to "π." The variation of the additional phase shift magnitude influences on the interference area size. Therefore, in framework of the waveguide-resonance model, the solution was accepted that the size of X-ray standing wave interference area is approximately conformed to half magnitude of the coherence length for the radiation flux undergoing the total reflection on the material interface [39].

If we place two planar dielectric polished reflectors on some distance, we can get the air planar extended slit clearance, which can be used for the realization of X-ray flux multiple external total reflection effects (**Figure 11a**). The consecutive multiple external total reflection phenomena are characterized by appearing the local interference area set. Since every elementary act of X-ray flux total reflection excites material volume of reflector, the next second reflection in the slit clearance on the reflector surface will lead to the material volume excitation. One can find specific X-ray flux incident angles, which will show the phasing of consecutive total reflection on every reflector (magic angles). Peculiarities of X-ray beam propagation through the air slit clearance are depended from the existence or absence of the phasing. The presence of phasing effect allows to transport the X-ray quasimonochromatic flux by PXW with small attenuation. The magic angle existence defines the discrete mode structure featured for the multiple total external reflection mechanisms. This picture is inherent for the X-ray polycapillary optics.

The mechanism of X-ray flux multiple total reflections is very efficient for the description of its propagation through planar wide slit clearances. But this mechanism is not able to characterize peculiarities of the radiation flux transportation by the super narrow planar extended slits. X-ray flux propagation through similar slits can be described on the basis of waveguide-resonance idea.

The conception of X-ray flux waveguide-resonance propagation is accompanied by appearing the X-ray standing wave uniform interference field in all narrow extended slit clearance spaces owing to the mutual overlap of local interference areas (**Figure 11b**). Overlay of these areas will be realized for any magnitudes of incident angles when it does not exceed the value of total reflection critical angle θ_c for the reflector material. The mode structure conception for PXWR is not existed. Moreover, we can confirm that the radiation coherence length magnitude is the critical parameter for the X-ray flux mechanism propagation change from the multiple external total reflections to the waveguide-resonance proceeding.

The narrow extended slit clearance radiation transport properties discussed above were investigated in the geometry when the projection of X-ray source focus was deposited in the symmetry plane of PXWR [39]. In this measurement geometry, the axis of X-ray incidence flux coincides with the axis of waveguide resonator.

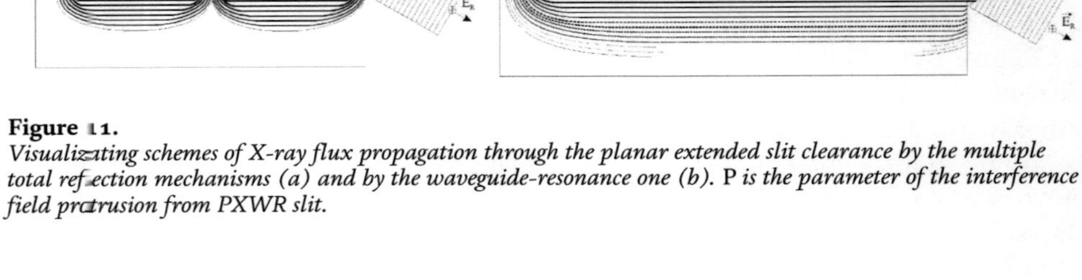

Figure 11.
Visualizating schemes of X-ray flux propagation through the planar extended slit clearance by the multiple total reflection mechanisms (a) and by the waveguide-resonance one (b). P *is the parameter of the interference field protrusion from PXWR slit.*

But in common case, the radiation flux can incident on the PXWR inlet at off-axis conditions. Measures in conditions of an initial X-ray flux off-axis incidence allow to differ the discrete mode structure from one with the continuous character, if the flux divergence is not great. The waveguide-resonance concept predicts that the off-axis incidence of X-ray flux will lead to the appearance of the emergent beam in the form of the double peak for a radiation spatial distribution (**Figure 11b**). The angular distance between the maximum of peaks must be equal to a double magnitude of the incidence angle. It is expected that the intensity of peaks must be approximately equivalent, and its divergences will be correlated with the radiation capture angle. Moreover, the integral intensity of the double-peak structure must be the monotonous function on the incidence angle for the angular interval $-\theta_c \leq \theta \leq \theta_c$. The reliable confirmation of all predictions following from the model of X-ray flux waveguide-resonance propagation was obtained in the course of our experimental investigations [36].

The integral intensity of PXWR emergent beam is insignificantly differed from the intensity of X-ray initial beam. Its magnitude can be described by the expression [39]:

$$W(x) = W_0 e^{-\alpha \mu x} \tag{4}$$

where W_0 is the initial beam intensity, μ is the linear absorption factor of reflector material for the radiation transported by PXWR, and α is the composite function defined by set factors of physical and geometrical nature. α magnitude is very small, and the attenuation of X-ray flux transported by PXWR is not significant. The experimental measurements showed that the attenuation factor for CuKα radiation transported by the quartz PXWR with a length of $l \sim 100$ mm can be near some per cents. The intensity losses for PXWR are smaller as the calculated values characterized for the multiple total reflection mechanisms. High radiation transportation efficiency of the planar X-ray waveguide-resonator is the result of the mode structure continuity for the flux propagating through its narrow extended slit clearance.

It is very important to notice that X-ray flux transportation by the waveguide-resonance mechanism is the result of the spatial coherence of quasimonochromatic radiation irradiated by X-ray tube. Owing to the fundamental physical reasons, a single X-ray photon in conditions of the total external reflection on the material interface cannot undergo interference with itself. The reflection process for X-ray and for other nature waves is accompanied by the Goos-Hanchen wave front displacement of the beam reflection position about the point of the beam incoming place, which is presented by the expression [42–44]:

$$\Delta x = \frac{\lambda_0}{\pi} \frac{1}{\sqrt{(\theta_c^2 - \theta^2) + 2i\beta}} \frac{1}{\sqrt{\theta_c^2 + 2i\beta}} \tag{5}$$

where θ_c is the magnitude of the total reflection critical angle. Minimum and maximum magnitudes of the displacement are arrived at $\theta = 0$ and $\theta = \theta_c$, correspondingly. The expressions for these values have forms:

$$\Delta x_{\min} = \frac{\lambda_0}{\pi \theta_c^2}; \quad \Delta x_{\max} = \frac{\lambda_0}{\pi \theta_c} \frac{1}{\sqrt{2\beta}} \tag{6}$$

Calculation of these factors shows that its magnitudes do not exceed the radiation coherence length and interference takes place.

6. Verification of the waveguide-resonance mechanism

The direct verification of the waveguide-resonance mechanism manifestation for the X-ray beam propagation can be found in the work of Japanese scientists [45]. The work was devoted to the transport property study of the angular structure as shown in **Figure 12a**. Japanese authors measured the MoKα flux intensity dependence on magnitude of the taper angle between two Si planar reflectors forming the radiation transportation structure. **Figure 12b** demonstrates the results of the measurements. The diffuse extremum *I* (near 0.1°) corresponds to reaching the critical total reflection angle for the molybdenum radiation on the silicon surface. The second extremum appearing near 0.007° cannot be explained without using the waveguide-resonance concept. The growth of X-ray radiation transport efficiency connected with this maximum reflects the transformation effect from the multiple total reflection propagations to the mechanism of the waveguide-resonance flux stream. The increase of an emergent beam intensity connects with a decrease in the flux attenuation featured for the waveguide-resonance propagation mechanism. The intensity of the second extremum is half of the first one, and the width of it is smaller than the first one on approximately one order. In result, the beam corresponding to the second radiation maximum will be characterized by the enhanced radiation density. The discussed results can be conceded as the independent confirmation of the waveguide-resonance mechanism objective reality for the quasimonochromatic X-ray flux propagation through the extended nanosize slit clearances.

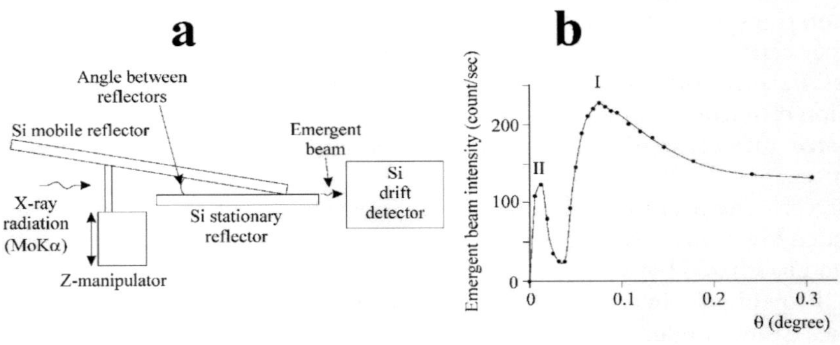

Figure 12.
Experimental scheme for the study of the radiation flux transporting peculiarities featured for the angular structure built on the basis of two Si reflectors under the variation of the taper angle between them (a), and the experimental diagram reflecting the emergent beam intensity dependence on the taper angle magnitude (b) [45].

7. Specific properties of PXWR

The waveguide-resonance mechanism is characterized by some specific properties of the quasimonochromatic radiation flux propagation through narrow extended slits, and the coherence length parameter is the limiting factor for the mechanism realization. The white radiation generated by X-ray tube is not characterized by parameter of the coherence length owing to the nature of this radiation arising [46]. But the experimental data presented in **Figure 13** show that the white radiation is transported by PXWR. At the same time, its related deposit at the total intensity of X-ray beam formed by PXWR is smaller than one in the beam formed by slit-cut system. So, one can expect that the spatial coherence degree for the white radiation generated by X-ray tube is smaller than this parameter featured for X-ray quasimonochromatic lines. **Figure 13** shows that the white component intensity falls down approximately two times in all spectral ranges investigated in the experiments. Thus, a planar X-ray waveguide resonator cannot be considered as a restrictive filter for the hard white radiation. But PXWR application for X-ray beam formation decreases the white radiation deposit in the total beam intensity. This effect will be greatest for the smallest slit clearance width. The specific feature of PXWR is the impossibility to use it for β-filtration of X-ray tube initial radiation. The β-filtration procedure for X-ray diffractometry is well known [47]. This procedure is based on the use of the thin film absorber manufactured from the material, which is characterized by the energy absorption edge intervened between EK_α and EK_β of the tube characteristic radiation. Similar β-filter can be built on the basis of planar monocapillar prepared by using the dielectric reflectors containing a significant concentration of atoms characterized by a suitable value of the energy absorption edge. Our direct experiments showed that the similar approach is not right for PXWR. β-Radiation flux excites the uniform interference field of X-ray standing wave in all space of PXWR air slit clearance, and the intensity attenuation is not observed.

Specific properties of PXWR are not exhausted by the peculiarities discussed above. For example, the beam formed by the waveguide resonator has the nanosize width and the enhanced radiation density. The beam is not accompanied by diffraction satellites and can be modulated by an external influence. But the planar X-ray waveguide resonator is characterized by two serious lacks. The angular divergence of the beam formed by PXWR of the simplest design is usually near 0.1°, and its real integral intensity is smaller than the integral intensities of beams formed by the slit-cut systems and the polycapillary optic devices on 1–2 orders [39]. The angular divergence of PXWR emergent beam can be decreased without influence on its integral intensity by application of PXWR with specific design, which has name as the composite planar X-ray waveguide resonator (CPXWR) [48].

Figure 14 presents the results of comparative investigations of X-ray characteristic beam formation by PXWR with the simplest construction (a) and CPXWR (b). Left part of the figure presents the measurement schemes. Spatial distributions of X-ray intensities in beams formed by these devices are shown in the right part of the figure. Radiation capture angle is the same and is equal to $\Delta\varphi_1 = 0.11°$. Composite

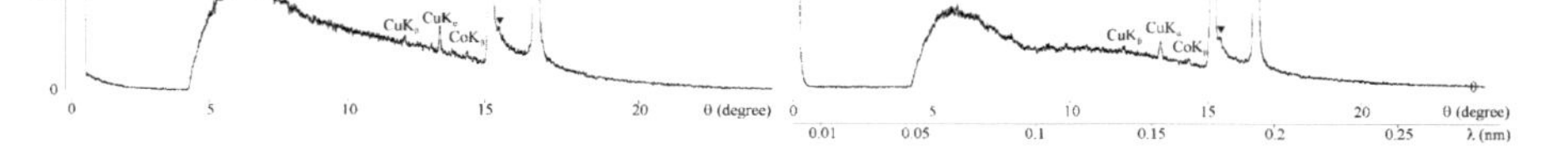

Figure 13.
Experimental diffraction patterns for SiO_2 (101) monocrystal specimen collected in conditions of a standard Bragg-Brentano geometry (a) and a waveguide-resonator application for the initial beam formation (b). The pattern normalization was carried out on the basis of equivalence of characteristic line intensities. Pattern (a) was registered at BSW-24 (Fe) X-ray tube regime U = 25 keV, I = 3 mA and pattern (b) U = 25 keV, I = 9 mA. Geometrical sizes at the measurements were (a) l_0 = 235 mm, l_1 = 50 mm, l_2 = 235 mm, $S_1 = S_2$ = 0.1 mm and (b) l_0 = 235 mm, l_1 = 50 mm, l_2 = 155 mm, l_3 = 85 mm, S_1 = 0.1 mm, s_{PXWR} = 0.1 µm. The collection was carried out without a pulse discrimination.

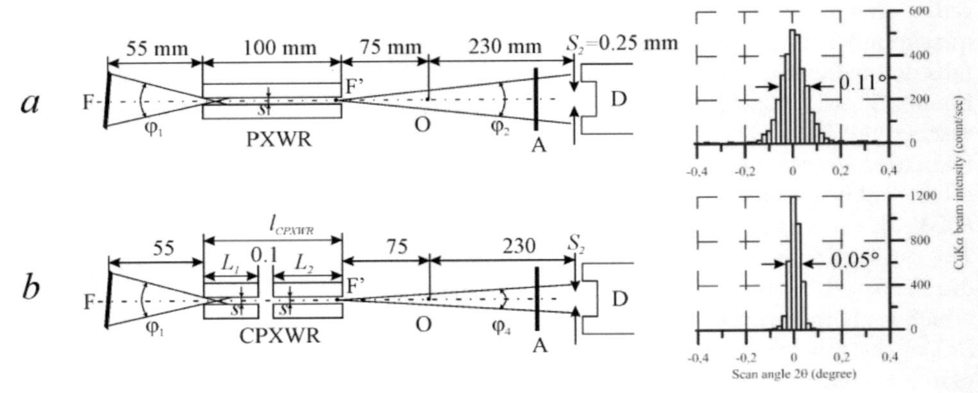

Figure 14.
Experimental schemes and flux intensity spatial distributions for CuKαβ beams formed by PXWR (a) and CPXWR (b) [48]. $S_{PXWR} = S_{CPXWR} = 88$ *nm,* $L_{PXWR} = L_{CPXWR} = 100$ *mm,* $l_1 = 75$ *mm,* $l_2 = 60$ *mm,* $l_3 = 400$ *mm,* $S_1 = 0.1$ *mm,* $L_1 = L_2 = 50$ *mm,* $\Delta L = 0.1$ *mm. Source regimes of BSW-24 (Cu) for both measurements* U = 20 *keV,* I = 10 *mA,* A *– filter attenuation factor* K = 200.

waveguide resonator differs from PXWR with the simplest construction by gap existence $\Delta L \sim 0.1$ mm between two PXWRs with the simplest construction built on the basis of short reflectors. The divergence of the waveguide emergent beams was studied by the method of the step-by-step detector scanning. The angular size of the detector slit S_1 was near 0.01°. The scanning step was $\Delta(2\theta) = 0.02°$.

Experimental intensity distributions for the beams formed by the conventional and the composite PXWRs demonstrate the Gauss form of distributions. FWHM of the peak distinguished for PXWR is $\Delta\varphi_2 = 0.11°$. At the same time, the magnitude of this parameter for CPXWR emergent beam is $\Delta\varphi_2 = 0.05°$ only. Total intensities of the peaks are approximately the same. Data presented show that the gap existence leads to the beam angular constriction without intensity losses. Such result is very alike on existence of the tunneling effect in the gap space. The increasing of ΔL distance up to 10 mm has led to an abrupt decrease of the peak total intensity and its FWHM.

Using the modified reflectors for the waveguide-resonator building allows to solve the second PXWR problem – low integral intensity of its emergent beam. Standard quartz glass plates modified by 30 mm polished tapers with an angle of $\psi = 0.5°$ were used for building the specific waveguide resonator (**Figure 15b**). For further radiation gathering power enhancement, the tapers were coated by HfO_2 thin film. Then, Ti strips with 90 nm thickness were deposited onto one plate edges, and the waveguide-resonance structure with a slit channel width of 90 nm and a height of 4 mm was assembled. In result, we received the skewed input concentrator with an angular aperture near 1°. Next, the comparative measurements of the conventional PXWR and the modified waveguide resonator were executed. The geometric parameters of the measuring schemes are given in **Figure 15**. Intensity spatial distributions for beams formed by the tested devices are shown in the same figure. In addition, the values of the total intensity (with and without the use of attenuator A) and the angular divergence of the beams are also quoted therein. The distributions were obtained at radiation source operation conditions [BSW-24 (Fe), $U = 20$ keV, $I = 10$ mA].

The data show that the envelope shape and FWHM of the intensity spatial distribution for a quasimonochromatic component of the beams formed by the conventional and the modified PXWR are nearly the same. On the other hand, the total intensity of the beam formed by the modified waveguide resonator is substantially higher than the beam intensity formed by the conventional PXWR. The data

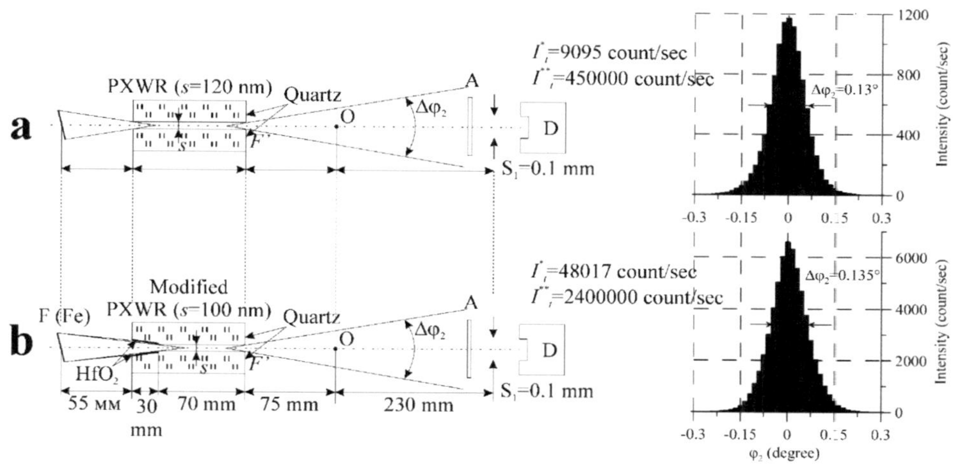

Figure 15.
Experimental schemes for the study of the intensity spatial distributions in X-ray beams formed by the waveguide resonator of a conventional construction (a) and PXWR with specific design (b) and real intensity distributions in its emergent beams. I^* *is the measurement magnitude, and* I^{**} *is the intensity corrected by taking into account the attenuator (A).*

show a fivefold increase in the radiation gathering power of the waveguide resonator due to the application of the input skewed radiation concentrator. Experimental value of the radiation gathering power enhancement obtained in our measurements was somewhat less than the rating. It is presumably explained by the nonoptimal length and form of the tapers. Nevertheless, the above result allows to state that the application of the input skewed concentrator is a powerful tool for the radiation gathering power enhancement of the waveguide-resonance structures, which provides the system modification without a significant loss in other parameters of the emergent beam.

In principle, there are other methods for the improvement of waveguide-resonator parameters. The most drastic method of PXWR parameter modification is the building of the multi-slit waveguide-resonance structures. According to our opinion, this way is very perspective, but it entails serious problems connected with the interference effect between the individual beams [49].

8. Practical application of the phenomenon

Experiments showed that the phenomenon of X-ray flux waveguide resonance increases the efficiency of X-ray fluorescence material analysis in conditions of exciting beam total reflection on studied surface (TXRF) [50]. This method modification by PXWR including the setup of TXRF spectrometer allowed to decrease the pollution detection limits in comparison with the convention of 1–2 orders. PXWR uses in experimental scheme of the particle induced X-ray emission (PIXE) allowed to elaborate the new experimental method for surface material element diagnostic [51]. Moreover, in some specific geometries, the method can provide element surface analysis being free from matrix effects.

The waveguide-resonance propagation of X-ray characteristic radiation fluxes can be achieved not only in frame of the external total reflection phenomenon but also at the use of the Bragg reflection. By using the Bragg-Laue waveguide-resonance cell (BLWRC), it is possible to build the pulsed X-ray laser on table, which will be useful for the study of kinetic processes [52]. Based on the use of phenomenon consequences, it is possible to realize the reactions of cold nuclear fusion [53]. But the more important result of waveguide-resonance radiation propagation phenomenon discovery, we regard the possibility appearing to elaborate the function correct model for optical fibers and waveguides of light beams. Conventional model of its function is based on the light flux notion as the infinite plane wave and on the light flux transportation mechanism by planar symmetrical waveguide as the multiple internal total reflections in frame of the geometrical paradigm [54–63]. Similar approach is not right, in principle. It is well known that any radiation source generates quasimonochromatic beams with λ_0 mean wavelength and $\Delta\lambda$ monochromatization degree. So, any quasimonochromatic beam is characterized by the coherence length parameter. Up-to-date optical lasers generate the beams with several tens of meters of coherence length. Owing to the core size of planar symmetrical optical waveguides varies from some micrometers to some millimeters, we can conclude that all optical waveguides and fibers are functioned in frame of the waveguide-resonance phenomenon manifestation, and instead of mode structure, it is a need to discuss the properties of uniform interference field of optical radiation standing wave.

9. Conclusion

The paper presented some experimental results allowed to consider that the waveguide-resonance conception is right. We described the features of the waveguide-resonance mechanism discussed its principle model and fixed the critical parameter being responsible for mechanism change of the radiation flux propagation. References presented allowed to understand some practical application of the phenomenon and its consequences. The paper contains the short description of X-ray device functioned on the basis of mechanism – the planar X-ray waveguide resonator, the discussion of PXWR properties, and the ways of its construction perfection.

Acknowledgements

The authors would like to thank Prof. J. Kawai and Prof. R. Van Grieken for the great attention to the waveguide-resonance direction of X-ray optics and Dr. M.S. Afanas'ev for help.

DOI: http://dx.doi.org/10.5772/intechopen.93174

Author details

Egorov Evgenii Vladimirovich[1,2,3]* and Egorov Vladimir Konstantinovich[1]

1 Institute of Microelectronics Technology Russian Academy of Science, Chernogolovka, Russia

2 Institute of Radio Engineering and Electronics Russian Academy of Science, Fryazino, Russia

3 Financial University under the Government of Russian Federation, Moscow, Russia

*Address all correspondence to: egorov@iptm.ru

References

[1] Hirsch P, Kellar J. An X-ray micro-beam technique: I-collimation. Proceedings of the Physical Society of London Series B. 1951;**64**:369-374

[2] Spiller E, Segmuller A. Propagation of X-rays in waveguide. Applied Physics Letters. 1974;**24**(2):60-61

[3] Feng Y, Sinha S, Fullerton E, Grubel G, Abernathy D, Siddons D, et al. X-ray Fraunhofer diffraction patterns from a thin film waveguides. Applied Physics Letters. 1995;**67**(24):3647-3649

[4] Lagomarsino S, Jark W, Di Fonzo S, Cedola A, Mueller B, Engstrom P, et al. Submicrometer X-ray beam production by a thin film waveguide. Applied Physics. 1996;**79**(8):4471-4473

[5] Jark W, Di Fonzo S, Lagomarsino S, Cedola A, Di Fabrizio E, Bram A, et al. Properties of a submicrometer X-ray beam at the exit of a waveguide. Applied Physics. 1996;**80**(9):4831-4836

[6] Marton J. The glass tube as X-ray guide. Applied Physics Letters. 1966;**9**(2):194-197

[7] Kumakhov M, Komarov F. Multiple reflection from surface X-ray optics. Physics Reports. 1990;**191**:289-352

[8] Vincze L, Janssens K, Adams F, Ridby A. Detail ray-tracing code for capillary optics. X-ray Spectrometry. 1995;**24**(1):27-37

[9] Arkadiev V, Beloglazov V, Bjeoumikhov A, Gorny H, Langhoff N, Wedell R. Application of capillary optics in modern scientific instrumentation. Poverkhnost. (X-ray, Synchrotron and Neutron Investigations). 2000;**1**:48-54 (in Russian)

[10] Haschke M, Haller M. Examination of poly-capillary lenses for their use in micro-XRF spectrometers. X-ray Spectrometry. 2003;**32**(3):239-247

[11] Rath B, Wang L, Homann B, Gibson W, MacDonald C. Measurements and analysis of radiation effects in polycapillary X-ray optics. Applied Physics. 1998;**83**(12):7424-7425

[12] Mingazin T, Zelenov V, Lejkin V. Slitless collimator for X-ray beams. Instruments and Experimental Techniques. 1981;**24**(1 part 2):244-247

[13] Leykin V, Mingazin T, Zelenov V. Collimating device for X-ray radiation. Pribori i Tekhnika Experimenta. 1981; **24**(3):208-211 (in Russian)

[14] Lejkin V, Mingazin T, Zelenov V. X-ray beam forming by using of a slitless collimator. Instruments and Experimental Techniques. 1984;**27**(6, part 1):1333-1336

[15] Egorov V, Zuev A, Maljukov B. Surface contamination diagnostics of silicon wafers by total reflection X-ray fluorescence spectrometry. Tsvetnaja metallurgija. Izvestija VUZoV. 1997;**5**: 54-69 (in Russian)

[16] Egorov V, Kondratiev O, Zuev A, Egorov E. The modification of TXRF method by use of X-ray slitless collimator. Advances in X-ray Analysis. 2000;**43**:406-417

[17] Zwanenburg M, Peters J, Bongaerts J, de Vries A, Abernathy D, van de Veen J. Propagation of X-ray in a planar waveguide with a turnable air gap. Physical Review Letters. 1999;**82**(8): 1696-1699

[18] Zwanenburg M, van de Veen J, Ficke H, Neerings H. A planar X-ray waveguide with a turnable air gap for the structural investigation of confined fluids. Review of Scientific Instruments. 2000;**71**(4):1723-1732

[19] Kawai J, Harada S, Karimov P. X-ray wave guide and its possible application

to surface analysis. Surface Analysis. 2002;**9**(3):356-358

[20] Pfeiffer T, Salditt P, Hoghoj F, Anderson I, Schell N. X-ray waveguides with multiple guiding layers. Physics Review. 2000;**B62**:16939-16943

[21] Ebel H, Svagera R, Ebel M. Multiple ionization in quantitative XRF. X-ray Spectrometry. 2001;**30**:180-185

[22] Dabagov S. Wave theory of X-ray scattering in capillary structures. X-ray Spectrometry. 2003;**32**:179-185

[23] Fuhse C, Salditt T. Finite-difference field calculations for one-dimensionally confined X-ray waveguides. Physica. 2005;**B357**:57-60

[24] Bukreeva J, Popov A, Pelliceia D, Cedola A, Dabagov S, Logomarsito S. Wave-field formation in a hollow X-ray waveguide. Physical Review Letters. 2006;**97**:184801-1(4)

[25] Pankin S, Hartman A, Salditt T. X-ray propagation in tapered waveguides: Simulation and optimization. Optic Communication. 2008;**281**:2779-2783

[26] Salditt T, Kruger S, Fuhse C, Bahtz C. High transmission planar X-ray waveguide. Physical Review Letters. 2008;**100**:184801-1(4)

[27] Blochin M, Sheitser I. X-ray Handbook. Moscow: Nauka; 1982. 375 p (in Russian)

[28] Zschornack G. Handbook of X-Ray Data. Berlin: Springer; 2007. 969 p

[29] Egorov V, Egorov E. Experimental study of X-ray the energy spectrum formed by a planar waveguide-resonator with specific element reflectors. Advances in X-Ray Analysis. 2006;**v49**:315

[30] Gasnier M, Nevot L. Analysis of crystallographic structures of chromium thin films. Physica Status Solidi. 1981; **A66**:525

[31] Harrick N. Internal Reflection Spectroscopy. New York: InterScience; 1967. 327 p

[32] Egorov V, Egorov E. Ion beam for materials analysis: Conventional and advanced approaches. In: Ahmad I, Maaza M, editors. Ion Beam Application. London: IntechOpen; 2018. 38 p

[33] Doolittle L. Algorithm for the rapid simulation of Rutherford backscattering spectra. Nuclear Instruments and Methods in Physics Research B. 1985;**9**: 344

[34] Turner AJ. Modern state of multilayer optical films field. Physique de Radium. 1950;**11**:444 (in French)

[35] Egorov V, Egorov E. Waveguide-resonance mechanism for X-ray beam propagation: physics and experimental background. Advances in X-ray Analysis. 2003;**46**:307

[36] Egorov V, Egorov E. Background of X-ray nanophotonics based on the planar air waveguide-resonator. X-ray Spectrometry. 2007;**36**:381

[37] Bohm M. Quantum Theory. New York: Prentice-Hall; 1952. 755 p

[38] Blochin M. Physik der Rontgenstrahlung. Berlin: Verlag; 1957. 535 p (in German)

[39] Egorov V, Egorov E. Planar waveguide-resonator: New device for X-ray optics. X-ray Spectrometry. 2004; **33**:360

[40] Landau L, Lifshitz E. Electrodynamics of Continues Medium. Oxford: Pergamon Press; 1984. 586 p

[41] Bedzyk M, Bommarito G, Schildkraud J. X-ray standing waves at a

reflecting mirror surface. Physical Review Letters. 1989;**69**:1376

[42] Goos F, Hanchen H. Novel and fundamental investigation of total internal reflection. Annalen der Physik. 1947;**6**(7–8, part 1):333 (in German)

[43] Snyder A, Love J. Goos-Hanchen shift. Applied Optics. 1976;**15**(1):236

[44] Brekhovskikh L. Waves in Lamellar Medium. New York: Academic Press; 1980. 345 p

[45] Tsuji L, Delalieus F. Characterization of X-ray emerging from between reflector and sample carrier in reflector-assisted TXRF analysis. X-ray Spectrometry. 2004;**33**: 281

[46] Compton A, Allison S. X-Ray in Theory and Experiment. New York: Princeton; 1935. 828 pp

[47] Klug H, Alexander L. X-Ray Diffraction Procedures. New York: Wiley; 1974. 976 pp.

[48] Egorov V, Egorov E. Peculiarities in the formation of X-ray fluxes by waveguide-resonators of different construction. Optics and Spectroscopy. 2018;**124**(6):838

[49] Mandel L, Wolf E. Optical Coherence and Quantum Optics. Cambridge: Cambridge University Press; 1995. 837 p

[50] Egorov V, Egorov E, Loukianchenko E. High effective TXRF spectrometry with waveguide-resonance devices application. Aspects in Mining and Mineral Science. 2018;**2**(4):1

[51] Egorov V, Egorov E, Afanas'ev M. High effective TXRF spectrometry with waveguide-resonance devices application. IOP Series: Journal of Physics: Conference Series. 2018;**1121**: 012011.

[52] Egorov V, Egorov E. Elaboration of pulsed X-ray laser on base of radiation fluxes waveguide-resonance propagation phenomenon. IOP Series: Journal of Physics: Conference Series. 2019;**1396**:012013

[53] Egorov V, Egorov E. About cold fusion possibilities in frame of the waveguide-resonance propagation of radiation fluxes. IOP Series: Journal of Physics: Conference Series. 2019;**1370**: 012021

[54] Kapany N. Fiber Optics: Principles and Application. New York: Academy Press; 1967. 429 p

[55] Shevchenko V. Continues Transition in Optics Waveguides. Boulder, Colorado: Golen Press; 1971. 176 p

[56] Kapany N, Burke J. Optical Waveguides. New York: Academic Press; 1972. 328 p

[57] Markuse D. Theory of Dielectric Optical Waveguides. New York: Academic Press; 1974. 256 p

[58] Unger H. Planar Optical Waveguides and Fibers. Oxford: Clarinton Press; 1977. 771 p

[59] Snyder A, Love J. Optical Waveguide Theory. London: Chapman and Hall Press; 1983. 734 p

[60] Buck J. Fundaments of Optical Fibers. 2nd ed. New York: Wiley; 2004. 352 p

[61] Okamoto K. Fundaments of Optical Waveguides. 2nd ed. New York: Academic Press; 2005. 584 p

[62] Calvo M, Lakshminarayanan V. Optical Waveguides, from Theory to Applied Technologies. New York: CRC Press; 2007. 401 p

[63] Kumar S, Deen M. Fiber Optic Communication. Chichester: Wiley; 2014. 573 p

Chapter 10

Characteristics of Radiation of a round Waveguide through a Flat Homogeneous Heat Shield

Viktor F. Mikhailov

Abstract

The problem of obtaining an analytical description of the radiation characteristics of a circular waveguide closed by a flat homogeneous dielectric plate is solved. The radiation characteristics include: the radiation field; the conductivity of the aperture radiation; and the fields of surface, flowing, and side waves, as well as energy characteristics. In such a statement, a strict solution of Maxwell's equations is required. The paper uses the method of integral transformations and the method of eigenfunctions. In this case, the assumption is used that the electrical parameters of the dielectric plate (thermal protection) and the geometric dimensions do not depend on time. The relations describing the directional diagram of a circular waveguide with dielectric thermal protection and taking into account the electrical parameters of thermal protection and its thickness are obtained. Expressions are also obtained for the fields of lateral, surface, and outflow waves, from which it is possible to calculate the power taken away by these fields. Numerical calculations were made for some of the obtained relations. The results showed that the power of the side waves is zero. It also follows from the calculations that the radiation field of surface and flowing waves is absent, that is, their contribution to the directional diagram is not.

Keywords: circular waveguide, a flat, uniform thermal protection, the radiation characteristics

1. Introduction

The onboard antennas of the returned spacecraft are subjected to intensive aerodynamic heating when the spacecraft passes through the dense layers of the atmosphere [1]. In these conditions, radio-transparent heat-resistant thermal protection is used to protect the antennas from external influences. The open ends of the transmission lines are used as the emitter to obtain a wide directional pattern. The most offer used radiation from the open end of the round waveguide. In the first approximation, we consider a flat uniform thermal protection. Under the conditions of aerodynamic heating, the electrical parameters of thermal protection significantly change (relative permittivity ε and tangent of the dielectric loss angle $tg\delta$). These changes lead to a noticeable increase in absorption losses in the heat shield, reflection from its boundaries, as well as to the appearance of surface and side waves. Together, all this leads to a change in the directional pattern and a decrease in the efficiency of the onboard antenna.

Evaluation of these changes is absolutely necessary to determine the radio technical characteristics of the onboard radio equipment. The paper solves the problem of determining the characteristics of the radiation of a circular waveguide through a heat shield subjected to aerodynamic heating.

The problems of calculating the interaction of the onboard antenna with a heat-shielding dielectric insert are very difficult and poorly developed. The development of mathematical models of antenna windows is reduced to solving an external problem of electrodynamics—electromagnetic excitation of bodies or diffraction of radio waves. At the same time, we will use well-known analytical methods of solution. The radio technical characteristics of the antenna window, for which we obtain an analytical description, include a radiation pattern, radiation conductivity, antenna temperature, and a number of other characteristics that describe more subtle electrodynamic effects, as well as energy characteristics.

In theoretical terms, the electrodynamic problem in general can be formulated as follows. There is a radiating, open antenna a, located on an infinite screen, in front of which is a dielectric layer of thickness d with a complex permittivity-bridge $\varepsilon\alpha(x, y, z, t)$ (see **Figure 1**).

In this general formulation, the solution of the electrodynamic problem is associated with significant mathematical difficulties, the main problem being the need to solve the Maxwell equations for an arbitrary law of change in the parameters of media in space and time. With some simplifying assumptions, the problem was solved in the ray approximation. In [2], the wave front method is used to analyze the radiation diagram of an antenna covered by a dielectric layer. In [3–5], the method proposed in [6] is used to find the radiation diagram, according to which the antenna radiation diagram in the presence of an infinite flat dielectric layer is simply multiplied by the diagram in the free space by the flat wave transmission coefficient for the flat layer, taking into account the corresponding angle of arrival and the plane of polarization of the wave.

At their core, all these methods are close to each other and are essentially based on the approximation of geometric optics, which is true in the quasi-optical domain. In relation to the problem under consideration (the resonance region), a strict solution of the Maxwell equations is required. From analytical methods of the solution, it is possible to apply the method of integral transformations and the method of eigenfunctions. Both of these methods will be used in the future. In this case, we assume that the parameters of thermal protection do not depend on time,

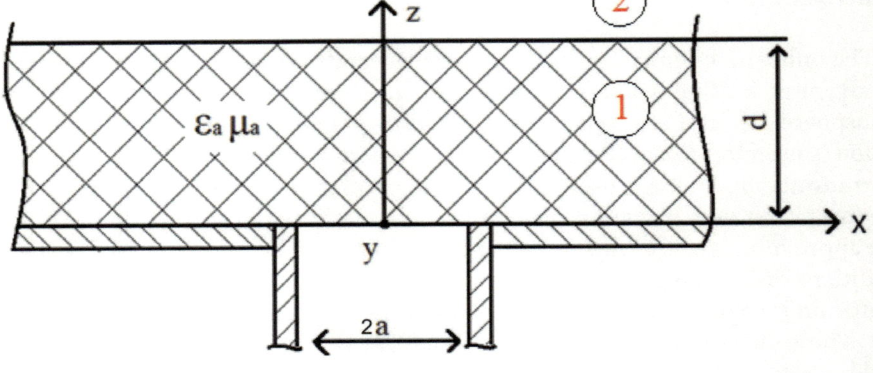

Figure 1.
Electrodynamic model of radiation from a circular waveguide with thermal protection.

that is, in fact, the slow-moving processes of heating of the dielectric thermal protection are considered. This approach and methods of solution were used in a number of works, for example [7].

2. Main part

The problem can be formulated as a boundary with respect to the tangent magnetic field in the material, and with respect to the tangent electric field. The second method is more convenient because of the simple form of boundary conditions for z = 0 [8].

The magnetic component of the electromagnetic field H_y at z ≥ 0 must satisfy the following wave equation in a Cartesian coordinate system x, y, z:

$$\frac{\partial^2 H_y}{\partial x^2}+\frac{\partial^2 H_y}{\partial y^2}+\frac{\partial^2 H_y}{\partial z^2}+k^2\varepsilon H_y=0, \tag{1}$$

where $\varepsilon = \varepsilon_1$ for $0 \le z \le d$, $\varepsilon = 1$ for z > d, and k is the wave number.

We apply the Fourier transform for x and y coordinates to equation (Eq. (1)). We get

$$\frac{\partial^2 \hat{H}_y}{\partial z^2}+\left(k^2\varepsilon-k^2{}_x-k^2{}_y\right)\hat{H}_y=0, \tag{2}$$

where is the direct Fourier transform,

$$\hat{H}_y=\int\int_{-\infty}^{\infty} H_y(x,y,0)\exp\left(-j\left(k_x x+k_y y\right)\right)dxdy.$$

Solution of equation (Eq. (2)) satisfying the radiation conditions (for $z \ge d$) has the form for area 1, that is, the area occupied by the dielectric plate (0 < z < d),

$$\hat{H}_y^{(1)}=D\exp\left(-jk_{z_1}z\right)+L\exp\left(jk_{z_1}z\right).$$

For region 2, that is, the region behind the plate (z ≥ d), we get

$$\hat{H}_y^{(2)}=M\exp\left(-jk_z z\right),$$

where $k_{z_1}=\sqrt{k^2\varepsilon-k_x^2-k_y^2}; k_z=\sqrt{k^2-k_x^2-k_y^2}.$

Reasoning in a similar way, with respect to the tangent magnetic component of the field, we obtain the following equations for the spectral component:

$$\hat{H}^{(1)}{}_x=A\exp\left(-jk_{z_1}z\right)+B\exp\left(jk_{z_1}z\right),$$

$$\hat{H}^{(2)}{}_x=C\exp\left(-jk_z z\right).$$

Satisfying the equations arising from the Maxwell equation, we obtain expressions for the spectral components of the electric field

$$\hat{E}_x^{(1)} = -\frac{k_{z_1}}{\omega\varepsilon_0\varepsilon_1}\left(Dexp\left(-jk_{z_1}z\right) - Lexp\left(jk_{z_1}z\right)\right);$$

$$\hat{E}_x^{(2)} = -\frac{k_z}{\omega\varepsilon_0}Mexp(-jk_z z);$$

$$\hat{E}_y^{(1)} = \frac{k_{z_1}}{\omega\varepsilon_0\varepsilon_1}\left(Aexp\left(-jk_{z_1}z\right) - Bexp\left(jk_{z_1}z\right)\right);$$

$$\hat{E}_y^{(2)} = \frac{k_z}{\omega\varepsilon_0}(Cexp(-jk_z z)).$$

Unknown functions A, B, C, D, L, and M are determined from the boundary conditions for z = 0 and z = d. In this case, the boundary conditions for radiation from a circular waveguide, when the field in the aperture is determined by H_{11} waves, have the form

$$-\frac{k_{z_1}}{\omega\varepsilon_0\varepsilon_1}(D-L) = \hat{E}_{x_0}; \quad (3)$$

$$\frac{k_{z_1}}{\omega\varepsilon_0\varepsilon_1}(A-B) = \hat{E}_{y_0}; \quad (4)$$

$$D\exp\left(-jk_{z_1}d\right) - L\exp\left(jk_{z_1}d\right) = \frac{\varepsilon_1 k_z}{k_{z_1}}M\exp(-jk_z d); \quad (5)$$

$$A\exp\left(-jk_{z_1}d\right) - B\exp\left(jk_{z_1}d\right) = \frac{\varepsilon_1 k_z}{k_{z_1}}C\exp(-jk_z d); \quad (6)$$

$$-k_xk_y\left(Dexp\left(-jk_{z_1}d\right) + Lexp\left(jk_{z_1}d\right)\right) + (k_1^2 - k_x^2)\left(Aexp\left(-jk_{z_1}d\right) + Bexp(jk_z d)\right) =$$
$$= \varepsilon_1((k^2 - k_x^2)Cexp(-jk_z d) - k_xk_yMexp(-jk_z d)); \quad (7)$$

$$\left(k_1^2 - k_y^2\right)\left(Dexp\left(-jk_{z_1}d\right) + Lexp\left(jk_{z_1}d\right)\right) - k_xk_y\left(Aexp\left(-jk_{z_1}d\right) + Bexp\left(jk_{z_1}d\right)\right) =$$
$$= \left(\left(k^2 - k_y^2\right)Mexp(-jk_z d) - k_xk_yCexp(-jk_z d)\right)\varepsilon_1 \quad (8)$$

where

$$\hat{E}_{x_0} = \iint_{\Pi} E_x(x', y', 0)\exp\left(-j(k_x x' + k_y y')\right)dx'dy'. \quad (9)$$

$$\hat{E}_{y_0} = \iint_{\Pi} E_y(x', y', 0)\exp\left(-j(k_x x' + k_y y')\right)dx'dy'. \quad (10)$$

Here, Π is the area of integration on the opening of the waveguide and x',y' are the coordinates counted in the opening of the studied waveguide.

We present in full the expression of the coefficients included in the solution of the wave equations of spectral components. In this case, we use the following variable replacement:

$$k_x = \beta\cos\alpha, \quad k_y = \beta\sin\alpha, k_z = \sqrt{k^2 - \beta^2}, \quad k_{z1} = \sqrt{k^2\varepsilon_1 - \beta^2}.$$

Since the coefficient expressions are very cumbersome, we use the following notation to obtain a compact form of the record:

$$\sqrt{k^2-\beta^2}=a,\quad \sqrt{k^2\varepsilon_1-\beta^2}=b,\quad \sqrt{k^2\varepsilon_1-\beta^2}\mathrm{d}=\mathrm{c},$$

$$\beta^2\sin\alpha\cos\alpha=\mathrm{e}, k^2-\beta^2\cos^2\alpha=\mathrm{f}, k^2\varepsilon_1-\beta^2\cos^2\alpha=\mathrm{g},$$

$$k^2-\beta^2\sin^2\alpha=\mathrm{h}, k^2\varepsilon_1-\beta^2\sin^2\alpha=\mathrm{l},\quad \sqrt{k^2-\beta^2}d=p.$$

As a result

$$\mathrm{A}=\frac{\omega\varepsilon_0\varepsilon_1}{b}\frac{\exp(jc)}{\Delta}\left[\hat{E}_{x_0}eabk^2(1-\varepsilon_1)(\cos c-j\sin c)+\right.$$
$$\left.+\hat{E}_{y_0}\left(e^2(a+b)(jb\sin c+a\cos c)+(bf+ag)(-j\sin ch\ nb\cos cla)\right)\right];\tag{11}$$

$$\mathrm{B}=\mathrm{A}-\hat{E}_{y_0}\frac{\omega\varepsilon_0\varepsilon_1}{b};\tag{12}$$

$$\mathrm{C}=\frac{b}{a}\exp\left[j(p+c)\right]\left[\hat{E}_{y_0}\frac{\omega}{b}-j\frac{2A}{\varepsilon_1}\sin c\exp(-jc)\right];\tag{13}$$

$$\mathrm{D}=\frac{\omega\varepsilon_0\varepsilon_1\exp(jc)}{b\Delta}\hat{E}_{x_0}\left[e^2(a+b)(jb\ \sin c+a\ \cos c)-(bh+al)\times\right.$$
$$\left.\times(jfb\sin c+ag\cos c)+\hat{E}_{y_0}aebk^2(1-\varepsilon_1)(j\sin c-\cos c)\right];\tag{14}$$

$$L=D+\hat{E}_{x_o}\frac{\omega\varepsilon_0\varepsilon_1}{b};\tag{15}$$

$$M=-\frac{b}{a}\exp\left[j(p+c)\right]\left[\hat{E}_{x_0}\frac{\omega\varepsilon_0}{b}+j\frac{2D}{\varepsilon_1}\sin c\ \exp(-jc)\right].\tag{16}$$

In the expressions (Eqs. (11)–(16))

$$\Delta=2\sqrt{k^2\varepsilon_1-\beta^2}\sqrt{k^2-\beta^2}k^2\left[j\sqrt{k^2-\beta^2}\sin\left(\sqrt{k^2\varepsilon_1-\beta^2}d\right)+\right.$$
$$+\sqrt{k^2\varepsilon_1-\beta^2}\cos\left(\sqrt{k^2\varepsilon_1-\beta^2}d\right)\left[\varepsilon_1\sqrt{k^2-\beta^2}\cos\left(\sqrt{k^2\varepsilon_1-\beta^2}d\right)+\right.$$
$$\left.\left.+j\sqrt{k^2\varepsilon_1-\beta^2}\sin\left(\sqrt{k^2\varepsilon_1-\beta\,d}\right)\right]\right..$$

Using the obtained expressions of the angular spectrum of plane waves and applying the inverse Fourier transform, taking into account (Eqs. (9) and (10)), we write

$$\begin{aligned}E_x^{(1,2)}=&\iint_{\Pi}F_{x_1}^{(1,2)}(x,y,z,x',y',0)E_x(x',y',0)dx'dy'+\\&+\iint_{\Pi}F_{x_2}^{(1,2)}(x,y,z,x',y',0)E_y(x',y',0)dx'dy';\end{aligned}\tag{17}$$

$$\begin{aligned}E_y^{(1,2)}=&\iint_{\Pi}F_{y_1}^{(1,2)}(x,y,z,x',y',0)E_x(x',y',0)dx'dy'+\\&+\iint_{\Pi}F_{y_2}^{(1,2)}(x,y,z,x',y',0)E_y(x',y',0)dx'dy';\end{aligned}\tag{18}$$

The functions F_x in these expressions are determined based on equations (Eqs. (11)–(16)). After a series of transformations, you can write

$$F^{(2)}_{x_1} = \frac{1}{4\pi^2}\int_{-\infty}^{\infty}\int_0^{2\pi}\left\{\frac{j2\,\sin\,\sqrt{k^2\varepsilon_1-\beta^2}d}{\Delta}\left(\left[-\beta^4\sin^2\alpha\,\cos^2\alpha\left(\sqrt{k^2-\beta^2}+\right.\right.\right.\right.$$
$$\left.+\sqrt{k^2\varepsilon_1-\beta^2}\right)\left(j\sqrt{k^2\varepsilon_1-\beta^2}\right)\sin\left(\sqrt{k^2\varepsilon_1-\beta^2 d}\right)+\sqrt{k^2-\beta^2}\cos\left(\sqrt{k^2\varepsilon_1-\beta^2 d}\right)\Big]+$$
$$+\left[\sqrt{k^2\varepsilon_1-\beta^2}(k^2-\beta^2\sin^2 a)+\sqrt{k^2-\beta^2}\left(k^2\varepsilon_1-\beta^2\sin^2 a\right)\right]\Big[j(k^2-\beta^2\cos^2 a)\sqrt{k^2\varepsilon_1-\beta^2}\times$$
$$\times\sin\left(\sqrt{k^2\varepsilon_1-\beta^2 d}\right)+(k^2\varepsilon_1-\beta^2\cos^2 a)\sqrt{k^2-\beta^2}\cos\sqrt{k^2\varepsilon_1-\beta^2 d}\Big]\Big)-1\Big\}\times$$
$$\times\exp\left[-j(kz-(k_z+k_{z1})d\right]\exp\left[-j\beta[(x'-x)\cos a+(y'-y)\sin a]\right]\beta\,d\beta\,da=$$
$$=\frac{1}{4\pi^2}\int_{-\infty}^{\infty}\int_0^{2\pi}\varphi^{(2)}_x\exp\left(-jk_z z\right)\exp\left[-j\beta[(x'-x)\cos a+(y'-y)\sin a]\right]\beta d\beta\,da; \tag{19}$$

$$F^{(2)}_{x_2} = \frac{1}{4\pi^2}\int_{-\infty}^{\infty}\int_0^{2\pi}\frac{j2\,\sin\left(\sqrt{k^2\varepsilon_1-\beta^2 d}\right)}{\Delta}\beta^2\sin\alpha\,\cos\alpha\left(\sqrt{k^2\varepsilon_1-\beta^2}k^2\times\right.$$
$$\times\sqrt{k^2-\beta^2}\exp\left(j\sqrt{k^2-\beta^2 d}\right)\exp\left[-j(k_z z-(k_z+k_{z_1})d)\right]\times \tag{20}$$
$$\times\exp\left[-j\beta[(x'-x)\cos a+(y'-y)\sin a]\right]\beta\,d\beta\,da=$$
$$=\frac{1}{4\pi^2}\int_{-\infty}^{\infty}\int_0^{2\pi}\varphi^{(2)}_y\exp\left(-jk_z z\right)\left[-j\beta[(x'-x)\cos a+(y'-y)\sin a]\right]\beta d\beta\,da;$$

$$F^{(2)}_{y_1} = \frac{1}{4\pi^2}\int_{-\infty}^{\infty}\int_0^{2\pi}\frac{j2\,\sin\left(\sqrt{k^2\varepsilon_1-\beta^2 d}\right)}{\Delta}\beta^2\sin^2\alpha\,\cos^2\alpha\left(\sqrt{k^2-\beta^2}\times\right.$$
$$\times\sqrt{k^2\varepsilon_1-\beta^2}k^2(1-\varepsilon_1)\cos\left(\sqrt{k^2\varepsilon_1-\beta^2 d}\right)-\mathrm{j}\,\sin\left(\sqrt{k^2\varepsilon_1-\beta^2 d}\right)\times \tag{21}$$
$$\times\exp\left[-j((k_z z)-(k_z+k_{z1})d)\right]\exp\left[-j\beta[(x'-x)\cos a+\right.$$
$$\left.+(y'-y)\sin a]\right]\beta\,d\beta\,da=\frac{1}{4\pi^2}\int_{-\infty}^{\infty}\int_0^{2\pi}\xi^{(2)}_x\exp\left(-jk_z z\right)\times$$
$$\times\exp\left[-j\beta[(x'-x)\cos a+(y'-y)\sin a]\right]\beta\,d\beta\,da;$$

$$F^{(2)}_{y_2} = \frac{1}{4\pi^2}\int_{-\infty}^{\infty}\int_0^{2\pi}\left\{\frac{j2\,\sin\left(\sqrt{k^2\varepsilon_1-\beta^2 d}\right)}{\Delta}\left(\left[\beta^4\sin^2\alpha\,\cos^2\alpha\left(\sqrt{k^2-\beta^2}+\right.\right.\right.\right.$$
$$\left.\left.+\sqrt{k^2\varepsilon_1-\beta^2}\right)\right]\left[\left(j\sqrt{k^2\varepsilon_1-\beta^2}\sin\left(\sqrt{k^2\varepsilon_1-\beta^2 d}\right)+\right.\right.$$
$$\left.+\sqrt{k^2-\beta^2}\left(k^2\varepsilon_1-\beta^2\cos^2 a\right)\right]\Big[-j\sqrt{k^2\varepsilon_1-\beta^2}(k^2\varepsilon_1-\beta^2\sin^2 a)\times$$
$$\times\sin\left(\sqrt{k^2\varepsilon_1-\beta^2 d}\right)-\sqrt{k^2-\beta^2}(k^2\varepsilon_1-\beta^2\sin^2 a)\times \tag{22}$$
$$\times\cos\left(\sqrt{k^2\varepsilon_1-\beta^2 d}\right)\Big]\Big)-1\Big\}\exp\left[-j(k_z z-(k_z+k_{z1})d)\right]\times$$
$$\times\exp\left[-j\beta[(x'-x)\cos a+(y'-y)\sin a]\right]\beta\,d\beta\,da=$$
$$=\frac{1}{4\pi^2}\int_{-\infty}^{\infty}\int_0^{2\pi}\xi^{(2)}_y\exp\left(-jk_z z\right)\exp\left[-j\beta[(x'-x)\cos a+\right.$$
$$\left.+(y'-y)\sin a]\right]\beta\,d\beta\,da.$$

The obtained expressions (Eqs. (19)–(22)) together with (Eqs. (17) and (18)) determine the radiation field of a circular waveguide with uniform thermal protection through the tangent components of the electric field in the opening of the waveguide.

To calculate the integrals (Eqs. (20) and (22)) by the saddle method, it should be taken into account that, when the integration contour is deformed, it is necessary to bypass the branch points of the integrand and that the saddle path intersects the poles of the integrand. We determine which branch points and poles of the integrands must be taken into account in the said approximation. Integrals (Eqs. (20) and (22)) can be written as

$$I = \int_{-\infty}^{\infty}\int_{0}^{2\pi} f(\beta)\exp\left(-j(d-z)\sqrt{k^2-\beta^2}\right)\exp\left(-j\beta\rho\cos\left(\alpha-\varphi_0\right)\right)d\beta d\alpha =$$

$$= 2\pi\int_{-\infty}^{\infty} f(\beta)\exp\left(j(d-z)\sqrt{k^2-\beta^2}\right)J_0(\beta\rho)d\beta.$$

where $\varphi_0 = arctg(x'-x)/(y'-y), \rho = \sqrt{(x'-x)^2+(y'-y)^2}$,
$J_0(\beta\rho)$ is the Bessel function.
We use the asymptotic expression of the Bessel function for large arguments

$$J_0(\beta\rho) \approx \sqrt{\frac{2\pi}{\pi\beta\rho}}\left(\exp\left(j(\beta\rho - \pi/4)\right) + \exp\left(-j(\beta\rho-\pi/4)\right)\right).$$

Then, passing to the plane of the complex angle τ by replacing β = ksinτ, we obtain

$$I = \int_{-\infty}^{\infty} Af(\beta)\exp\left(jkL\cos\left(\tau\pm\varphi\right)\right)d\tau \tag{23}$$

where $A = 2k\sqrt{\frac{2\pi}{\beta\rho}}\cos\tau\exp\left(-j\pi/4\right), d-z = L\cos\varphi, \rho = L\sin\varphi$.

Considering the obtained expression of the integral I, we find the analytical expression of the saddle path from the following equation:

$J_m(\ j\cos\left(\tau\pm\varphi\right)) = \text{const}, \text{sciliset}\, J_m(\ j\cos\left(\tau\pm\varphi\right)) = j.$ Means $\cos\left(\tau\pm\varphi\right) = 1.$.
Because the$\tau = \tau_r + j\tau_j$,

then $\cos\left(\tau_r\pm\varphi\right) = \cos\left(\tau_r\pm\varphi\right)\text{ch}\tau_j - j\sin\left(\tau_r\pm\varphi\right)\text{sh}\tau_j$, $\cos\left(\tau_r\pm\varphi\right) = \frac{1}{ch\tau_j} =$ $\text{sch}\tau_j$.

Finally, we get the expression of the saddle path

$$\tau_r \pm \varphi = \arccos\left(sch\tau_j\right). \tag{24}$$

Satisfying the radiation condition at infinity, we obtain regions on the complex plane τ in which the saddle path defined by (Eq. (24)) lies:

1. $0 < \tau_r < \pi,\ \tau_i > 0$;

2. $-\pi < \tau_r < 0, \tau_j < 0$;

moreover, the second area of determination of the transit path in our case makes sense, since from (Eqs. (20) and (22)) it is clear that the lower limit of β is 0. Denoting the coordinates of the branch points τ_B and the pole τ_p, we obtain from

(Eq. (24)) the condition that determines the branch points and poles intersected during deformation of the initial path of integration into the saddle,

$$\tau_{z_в, z_p} \pm \varphi > \arccos\left(\mathrm{sch}\tau_{\ j_в, j_p}\right).$$

In the general case, in accordance with the Cauchy theorem, the integrals for functions can be represented in the following form:

$$F = \frac{1}{4\pi^2}\left(\int\limits_{l} \dots d\beta + U(C_в)\int\limits_{l_в} \dots d\beta + U(C_p)\int\limits_{l_p} \dots d\beta\right), \tag{25}$$

where $U(C_{в,\ p})$is the only Heaviside function; $C_в, C_p$ are values determined on the basis of Eq. (24) as follows on the complex plane:

$$C_{в,\ p} = Re\left(\arcsin\frac{\beta_{в,\ p}}{k}\right) \pm \varphi - \arccos\left(sch\, J_m\left(\arcsin\frac{\beta_{в,\ p}}{k}\right)\right), \tag{26}$$

The first integrals over the circuit l are calculated by the saddle method. Finally, by the saddle method, we get

$$F^{(2)}_{x1\ sad} = \frac{jk\ \exp(jkr)}{2\pi r^2} z\varphi_x^{(2)} \begin{vmatrix} k_x = k_{x'} \\ k_y = k_{y'} \end{vmatrix}, \tag{27}$$

$$F^{(2)}_{x2\ sad} = \frac{jk\ \exp(jkr)}{2\pi r^2} z\varphi_y^{(2)} \begin{vmatrix} k_x = k_{x'} \\ k_y = k_{y'} \end{vmatrix}, \tag{28}$$

$$F^{(2)}_{y1\ sad} = \frac{jk\ \exp(jkr)}{2\pi r^2} z\xi_y^{(2)} \begin{vmatrix} k_x = k_{x'} \\ k_y = k_{y'} \end{vmatrix}, \tag{29}$$

$$F^{(2)}_{y2\ sad} = \frac{jk\ \exp(jkr)}{2\pi r^2} z\xi_y^{(2)} \begin{vmatrix} k_x = k_{x'} \\ k_y = k_{y} \end{vmatrix}. \tag{30}$$

In the expressions (Eqs. (27)–(30))

$$k_x' = \frac{k_x}{r},\ k_y' = \frac{k_y}{r},\ r = \sqrt{(x'-x)^2 + (y'-y)^2 + z^2}.$$

After a series of transformations for even E modes and odd H modes, we obtain

$$F^{(2)}_{x_{1E}} = \frac{i}{2\pi}\sum_{i=1}^{n} \frac{U(C_{p_i})\int_0^{2\pi}\varphi_x^2(\beta_i, a)N\ (\beta_i, a)da\psi_1(\beta_i)}{\psi_1{}'(\beta_i)} \quad |\ \beta_i = \beta_i^E, \tag{31}$$

$$F^{(2)}_{x_{2E}} = \frac{i}{2\pi}\sum_{i=1}^{n} \frac{U(C_{p_i})\int_0^{2\pi}\varphi_y^2(\beta_i, a)N\ (\beta_i, a)da\psi_1(\beta_i)}{\psi_1{}'(\beta_i)} \quad |\ \beta_i = \beta_i^E, \tag{32}$$

$$F^{(2)}_{y_{1E}} = \frac{i}{2\pi}\sum_{i=1}^{n} \frac{U(C_{p_i})\int_0^{2\pi}\varphi_x^2(\beta_i, a)N\ (\beta_i, a)da\psi_1(\beta_i)}{\psi'(\beta_i)} \quad |\ \beta_i = \beta_i^E, \tag{33}$$

$$F^{(2)}_{y_{1E}} = \frac{i}{2\pi}\sum_{i=1}^{n} \frac{U(C_{p_i})\int_0^{2\pi}\varphi_y^2(\beta_i, a)N\ (\beta_i, a)da\psi_1(\beta_i)}{\psi_1{}'(\beta_i)} \quad |\ \beta_i = \beta_i^E. \tag{34}$$

In these expressions

$$N = -2\sqrt{k^2\varepsilon_1 - \beta^2}\sqrt{k^2 - \beta^2}k^2 \exp\left(-jk_z z\right)\times$$

$$\times \exp\left[-j\beta[(x' - x)\cos a + (y' - y)\sin a]\right]\beta;$$

$$\psi_1(\beta) = j\, ctg\left(\sqrt{k^2\varepsilon_1 - \beta^2 d}\right)k^2\varepsilon_1\sqrt{k^2 - \beta^2} - k^2\sqrt{k^2\varepsilon_1 - \beta^2},$$

C_{pi} is calculated according to (Eq. (26)). The expressions for the H modes will be characterized by equations similar to equations (Eqs. (31)–(34)), in which ψ_1 (β) is replaced by ψ_2 (β) and ψ_1 (β) by ψ_2 (β) and which will be calculated for the values of β corresponding to the poles, and

$$\psi_2(\beta) = \sqrt{k^2 - \beta^2} - j\sqrt{k^2k^2\varepsilon_1 - \beta^2}ctg(\sqrt{k^2\varepsilon_1 - \beta^2 d}.$$

Expressions $F^{(1)}_{x_{1E}}$, $F^{(1)}_{x_{2E}}$, $F^{(1)}_{x_{yE}}$, $F^{(1)}_{y_{1E}}$ are written similar to expressions with replacement, $\varphi^{(2)}_{xN}$ by $\varphi^{(1)}_{xN}$, $\varphi^{(2)}_{y_N}$ by $\varphi^{(1)}_{y}$, $\xi^{(2)}_{xN}$ by $\xi^{(1)}_{xN}$, and $\xi^{(2)}_{y_N}$ by $\xi^{(1)}_{y}$, respectively.
where

$$\varphi^{(1)}_{xN} = \varphi^{(1)}_{x(-)}\left[N_{(-)} - N_{(+)}\right] - N_{(+)}; \tag{35}$$

$$\varphi^{(1)}_{y_N} = \varphi^{(1)}_{y_{(-)}}\left[N_{(-)} - N_{(+)}\right]; \tag{36}$$

$$\xi^{(1)}_{xN} = \xi^{(1)}_{x_{(-)}}\left[N_{(-)} - N_{(+)}\right]; \tag{37}$$

$$\xi^{(1)}_{y_N} = \xi^{(1)}_{x_{(-)}}\left[N_{(-)} - N_{(+)}\right] - N_{(+)}. \tag{38}$$

In expressions (Eqs. (35)–(38)), the expression N (−) corresponds to the expression N with the replacement of exp ($-jk_z z$) by exp ($-jk_1 z$);
expression N (+) corresponds to expression N replacing exp ($-jk_z z$) with exp ($-jk_1 z$),

$$\varphi^{(1)}_{x_{(-)}} = \frac{\exp(jc)}{\Delta}\left[e^2(a+b)(jb\sin c + a\cos c) - (bh + al)(jb\sin c + ag\cos c)\right];$$

$$\varphi^{(1)}_{(-)} = \xi^{(1)}_{x(-)}\frac{aebk^2(\varepsilon_1 - 1)}{\Delta};$$

$$\xi^{(1)}_{y(-)} = \frac{\exp(jc)}{\Delta}\left[(bf + ag)(jb\sin c + al\cos c) - l^2(a+b)(jb\sin c + a\cos c)\right].$$

The relation for functions includes expressions of spectral components. Considering the main type of oscillations in the waveguide H_{11}, we get

$$\hat{E}_{x1} = \pi\omega\mu_0 H_0 a^6 k_x k_y\left[0,1 + 6,5 + 10^3 a^2\left(3,4 + k_x^2 + k_y^2\right) + 1,45\cdot 10^3 a^4\left(k_x^2 + k_y^2\right)\right];$$

$$\hat{E}_{y1} = 0,85\pi\omega\mu_0 H_0 a^2[1 - 1,25a^2\left(b^2 + k_x^2 + 0,5k_y^2 + 3,125\cdot 10^{-2}\right)\times$$

$$\times a^4\left(0,33\,k_x^2k_y^2 + 0,33b^2k_x^2 + 0,5b^2k_y^2\right) - 0,5\cdot 10^{-4}a^6b^2k_x^2k_y^2];$$

The second integral (Eq. (25)) is calculated along the contour l_B, which covers the cut and is carried out so that the integrand is unique. From the form of the integrands, it is obvious where the first-order branch points are located. From the form of the integrands of the integrals (Eq. (25)) it is obvious that the first-order branch points are located at $\beta_{\theta_1} = \pm k$ and $\beta_{\theta_2} = \pm k\sqrt{\varepsilon_1}$. From these branch points, it is necessary to take $\beta_{\theta_1} = k$ and $\beta_{\theta_2} = k\sqrt{\varepsilon_1}$, in order to satisfy the radiation conditions. A calculation according to (Eq. (26)) shows that $C_B > 0$ for named branch points and, with the exception of the case when ε1 is complex and the losses are sufficiently large (tanδ > o.5). Thus, we find that the side wave can contribute to the radiation field, and this should be taken into account. An analysis of the integrands (Eqs. (20) and (22)) showed that the developed and well-known methods for the asymptotic estimation of integrals along the banks of a section covering branch points that are valid for a saddle point turn out to be inapplicable in this case. For this reason, the contribution of the side wave to the radiation diagram can be determined only by numerical integration (Eq. (25)) along the contour l_B. Moreover, from the analysis of integrands, it follows that it is advisable to choose the section so that it is a straight line parallel to the imaginary axis of the complex plane β. Then the integral along the banks of the section $\int\limits_{l_\theta} ... d\beta$ will take the form

$$\int\limits_{\mathrm{Re}\,k+j\infty}^{k} v_1(\beta)d\beta + \int\limits_{k}^{\mathrm{Re}\,k+j\infty} v_2(\beta)d\beta,$$

where $v_1(\beta)$ and $v_2(\beta)$ are sub-integral expressions with signs in front of $\sqrt{k^2 - \beta^2}$.

The conditions for the existence of surface and leaky waves are determined from the location of the poles of the integrands (Eqs. (20) and (22)), and the poles correspond to the equality of the denominators of the marked expressions to zero.

By the Cauchy theorem, integral (Eq. (25)) along the contour lp is defined as follows:

$$\int\limits_{l_p} ... d\beta = \sum_{i=1}^{n} \mathrm{Re}\,s(\beta_i),$$

where β_i are the roots of the denominators (Eq. (25)).

In order to separate the singular points into poles corresponding to surface and outgoing waves, it is advisable to again go from the complex plane β to the plane of the complex angle τ. Moreover, from the analysis of the exponent of expression (23), it follows that the surface wave will take place at

$$\tau_r = {}^{\pi}\!/_{2}, \tau_j > 0$$

and the outgoing wave corresponds to the following region of complex angles:

$$0 < \tau < \pi, \text{(except for } \tau r = \pi/.2), \tau ji > 0.$$

Moreover, the region 0 < τr<ϕ determines the backward wave that does not satisfy the conditions at infinity.

In the presence of losses in the dielectric plate k1 = kr-jkj, analysis of the exponent (Eq. (23)) shows that the poles corresponding to the relations for the complex angle τ

$$\tau_r = \arccos\left(-\frac{\varepsilon tg\delta}{sh\tau_j}\right), \varepsilon - \sqrt{1 - \frac{\varepsilon^2 tg^2\delta}{sh^2\tau_j} ch\tau_j} < 0$$

determine the surface wave field.
If

$$\cos\tau_r sh\tau_j \neq -\varepsilon tg\delta,$$

then the poles that satisfy the last relation determine the field of the leaky wave. The conductivity of the aperture is determined by the following expression:

$$Y_a = \int\int_{-\infty}^{\infty} E_x^{(1)^*} H_y^{(1)} dx\, dy.$$

Applying the Parseval theorem, after a series of transformations, we obtain the expression for the conductivity referred to as the conductivity of the open end of the waveguide in the form

$$\begin{aligned} Y_{an} = {} & \frac{\omega\varepsilon_0}{2\pi^2 a}\int_0^{\infty}\int_0^{\pi} \frac{\beta\hat{E}_{x_0}^2 \exp(jc)}{\Delta}\left(\left(e\hat{E}_{x_0} + e\hat{E}_{y_0} eabk^2(1-\varepsilon_1)\times\right.\right. \\ & \times(\cos c - j\sin c) - \left(e\hat{E}_{x_0} - e\hat{E}_{y_0}\right)e^2(a+b)(jb\sin c + a\cos c) + \\ & + \frac{\Delta}{2\exp(jc)} + e\hat{E}_{x_0}(hb + la)(jfb\sin c + ag\cos c) - \\ & \left.\left. - e\hat{E}_{y0}(bf + ag)\right)d\beta\, da = \frac{\omega\hat{E}_0}{2\pi^2 b}\int_0^{\infty}\int_0^{2\pi} \frac{G(\beta a)}{\Delta} d\beta\, da. \end{aligned} \tag{39}$$

The contribution of surface and leaky waves to the conductivity of a circular waveguide with thermal protection is determined using the Cauchy theorem. As a result, we get

$$Y_{sur} = \frac{j\omega\varepsilon_0}{2\pi^2 a}\sum_{i=1}^{n}\int_0^{2\pi} \frac{G(\beta_i, \alpha)}{\Delta} d\alpha. \tag{40}$$

We will conduct a quantitative assessment of various loss mechanisms by means of transmission coefficients χ, attenuation υ, and reflection ρ, which are defined as follows:

$$\begin{aligned} \eta &= \eta_1 + \eta_2 + \eta_3 + \eta_4 = \frac{P_{rad}}{P_f} + \frac{P_{rad.sur}}{P_f} + \frac{P_{rad.res}}{P_f} + \frac{P_{rad.sid}}{P_f}, \\ \nu &= \nu_1 + \nu_2 + \nu_3 + \nu_4 = \frac{P_{rad} - P_{rad.sur}}{P_f} + \frac{P_{rad} - P_{rad.res}}{P_f} + \frac{P_{rad} - P_{rad.sid}}{P_f}, \\ |R| &= \left|\frac{1 - Y_n}{1 + Y_n}\right|. \end{aligned} \tag{41}$$

These expressions take into account the orthogonality of surface, leaky, and side waves and P_f is the power incident (supplied to the emitter), P_{rad}-radiated power,

$P_{rad.sur}$, $P_{rad.res}$, $P_{rad.sid}$. waves, respectively, P_{sur}, P_{res}, and P_{sid} are the power of surface, leaky, and side waves, respectively; υ_1 is the attenuation coefficient of radiated power, determined by losses in dielectric thermal protection.

$$\nu_1 = 1 - \eta - \nu_2 - \nu_3 - \nu_4 - |R|^2.$$

In expression (Eq. (41))

$$Y_n = Y_{an} - Y_{sur}.$$

3. Conclusion

Expressions (Eqs. (17), (18), (27)–(30), (39), and (40)) characterize the radiation field of a circular waveguide with uniform thermal protection and its input conductivity. The theoretical relations obtained, along with an estimate of the apparent loss of radiated power due to absorption in and reflection from thermal protection, allow us to estimate more subtle effects, such as losses on surface, leaky, and side waves. It is also possible to assess the influence of these waves on the radiation pattern.

Further development of the obtained theoretical relations should consist of taking into account the probable inhomogeneity of the heat-shielding layer in the direction of the z axis. For this, the method of approximate solution of the wave equation can be applied—the WKB method (Wentzel, Kramer, Bruellen) [9]. If we consider a round1 waveguide in the form of an onboard antenna of the returned spacecraft, then due to aerodynamic heating, a melt layer appears on the outer surface of the heat shield, which has electrical characteristics different from the characteristics of the material in the solid phase. Then the radiation from the waveguide should be considered through a two-layer dielectric thermal protection [10].

Unfortunately, all the obtained relations turned out to be very cumbersome and their use becomes only with numerical integration. For some of the obtained ratios, numerical calculations were performed. The results showed that the power of the side waves is zero. It also follows from the above calculations that the radiation field of surface and leaky waves is absent, that is, their contribution to the radiation pattern is not. Further research in this area should be directed to the development of computer calculation programs for the basic radiation characteristics.

Author details

Viktor F. Mikhailov
State University of Aerospace Instrumentation, St. Petersburg, Russia

*Address all correspondence to: vmikhailov@pochta.tvoe.tv

References

[1] Martin C, Baltimore M. Re-entry temperature error analyses; 1967. p. 69. DOI: 10.2172/4657807

[2] Hubregt JV. Antenna Theory and Applications. Vol. 265. John Wiley and Sons; 2012. DOI: 10.1002/9781119944751

[3] Matsumoto M, Tsutsumi M, Kumagai N. Millimetre-wave radiation characteristics of a periodically plasma-induced semiconductor waveguide. Electronics Letters. 1986;**22**(13):710-711. DOI: 10.1049/el:19860486

[4] Bachynski MP, Clouter RF. Free-space microwave measurements of plasma properties. Antennas and Propagation. 1963;**1**:269. DOI: 10.1109/APS.1963.1148676

[5] Burberry RA. VHF and UHF antennas. Institution of Engineering and Technology. 1992. p. 312. DOI: 10.1049/PBEW035E

[6] Richmond IN. Antenna pattern distortion by dielectric sheath. Trans. JRE. 1956;**AP-4**(2):43-51. DOI: 10.1109/TAP.1956.1144381

[7] Croswell WF, Rudduck RC, Hatcher D. The admittance of a rectangular waveguide radiating into a dielectric slab. Antennas and Propagation. 1967;**AP-15**(5):627-633. DOI: 10.1109/tap.1967.1139026

[8] Felsen L, Markuvits N. Radiation and scattering of waves. M .: Mir. 1978;**1**:547. DOI: 10.1109/9780470546307

[9] Vasilieva OV, Mikhailov VF. Radiation of a flat waveguide closed by inhomogeneous thermal protection. In: Proceedings Article published Jun 2019 in 2019 Wave Electronics and its Application in Information and Telecommunication Systems (WECONF). DOI: 10.1109/weconf.2019.8840596

[10] Mikhailov VF. Radiation of a flat waveguide closed by molted heat protection. In: Proceedings Article published Jun 2019 in 2019 Wave Electronics and its Application in Information and Telecommunication Systems (WECONF). DOI: 10.1109/weconf.2019.8840592